CO_2 Storage Coupled with Enhanced Oil Recovery

Kun Sang Lee · Jinhyung Cho · Ji Ho Lee

CO_2 Storage Coupled with Enhanced Oil Recovery

Springer

Kun Sang Lee
Department of Earth Resources
and Environmental Engineering
Hanyang University
Seoul, Korea (Republic of)

Jinhyung Cho
Department of Earth Resources
and Environmental Engineering
Hanyang University
Seoul, Korea (Republic of)

Ji Ho Lee
E&P Technology Center
Korea National Oil Corporation
Ulsan, Korea (Republic of)

ISBN 978-3-030-41900-4 ISBN 978-3-030-41901-1 (eBook)
https://doi.org/10.1007/978-3-030-41901-1

This Springer imprint is published by the registered company Springer Nature Switzerland AG
The registered company address is: Gewerbestrasse 11, 6330 Cham, Switzerland

Contents

Chapter 1
Introduction

Abstract A number of methods have been developed for reducing CO_2 emission over the last decades. Among them, interest in coupled CO_2 enhanced oil recovery (EOR) and carbon capture and storage (CCS) has increased because of its simultaneous consideration of business and climate change mitigation. When only CCS is applied into a deep saline aquifer, the consideration of two-phase (gas/water) system is enough to evaluate the CCS performance. However, for CO_2 injection into a mature oil reservoir for CCS and EOR, the three-phase (oil/water/gas) system should be considered. Since the injected CO_2 contacts and changes the composition and properties of oil causing asphaltene precipitation, four-phase (oil/water/gas/solid) system modeling is necessary for more precise prediction on the CCS and EOR performances. In this chapter, the various factors such as minimum miscibility pressure (MMP) and injection schemes and main mechanisms for CO_2 EOR are introduced briefly. It also explains the present status of coupled CCS and CO_2 EOR as a technology of carbon capture, utilization, and storage (CCUS) and asphaltene precipitation which affects its performance. Details of the technology will be explained in the following chapters.

Keywords CO_2 enhanced oil recovery (EOR) · Carbon capture · Utilization · And storage (CCUS) · Asphaltene

1.1 CO_2 Enhanced Oil Recovery

Oil production is generally conducted in several production phases, which are often grouped in three main stages. Firstly, primary oil production phase refers to oil flows to the production well as a result of pressure gradients in the reservoir. It is called the natural drive because this initial flow occurs naturally. The oil production rate tends to decrease over time because reservoir pressure declines and water production increases at the same time. At this time, secondary oil recovery is applied because primary production is not effective anymore.

The main purpose of secondary recovery is increasing reservoir pressure by injecting a fluid such as water or natural gas. The difference with tertiary recovery is that the injected fluid does not change the properties of oil and maintains the reservoir

K. S. Lee et al., *CO_2 Storage Coupled with Enhanced Oil Recovery*,
https://doi.org/10.1007/978-3-030-41901-1_1

pressure. Natural gas, which is produced during primary production, is usually re-injected into the gas cap to drive oil downward to the producer with high pressure because gas is much lighter than oil. Though the purpose of water flooding is similar to natural gas drive, water is typically injected into the other side of the oil reservoir to sweep oil directly toward the production well. Primary and secondary recovery can usually recover around 1/3 of original oil in place.

After secondary recovery reaches the time of no longer cost effective, enhanced oil recovery (EOR) is considered as a tertiary recovery method. EOR refers to the method to produce oil by the injection of fluids which do not originally exist in the reservoir, and it aims to change the oil properties. Gas, chemical, and thermal injections are major methods for EOR, and the appropriate method can be chosen depending on the type of oil and reservoir.

CO_2 EOR is considered as one of the most promising EOR methods due to the physical and chemical properties of the CO_2-oil system, CO_2 availability, and status of technology developments (Kamali et al. 2017). The utilization of CO_2 to enhance oil production has several decades of history, and most of this practice has been developed in North America (Gozalpour et al. 2005; IEA 2015a; Elmabrouk et al. 2017). When CO_2 is injected into reservoirs, it reduces oil viscosity, interfacial tension (IFT), and oil density, so it becomes mutually soluble with the residual oil as light to medium hydrocarbons. These effects occur under miscible condition more strongly than under immiscible condition (Makimura et al. 2013). Miscibility between oil and CO_2 depends on pressure, and the reservoir pressure should be maintained above the minimum miscibility pressure (MMP) to reach miscible condition through first or multiple contact miscibility process. First contact miscibility process occurs when injected fluid and reservoir oil contact at the first time with much higher reservoir pressure than MMP, and multiple contact miscibility process is attained by condensing and vaporizing gas drive mechanisms (Stalkup 1987; Johns et al. 1993; Bahadori 2018). However, CO_2 cannot be injected at unlimited high pressure because it may lead to fractures if it exceeds the reservoir fracture pressure. When reservoir pressure is maintained between MMP and fracture pressure, the injected CO_2 and reservoir oil are completely miscible as single phase. If reservoir pressure cannot reach MMP, immiscible flood is operated. The mechanisms of immiscible flood are similar to miscible flood, such as oil viscosity and density reduction, but the effects are smaller than those of miscible flood. Therefore, main force of immiscible flood is that the injected CO_2 physically pushes oil toward the producer as secondary recovery.

Several CO_2 EOR methods, such as continuous CO_2 flooding, huff-n-puff, and CO_2 water alternating gas (WAG), have been developed and applied (Han and Gu 2015). Though having been verified to be an effective EOR method by field cases, continuous CO_2 flooding has some limitations such as a low volumetric sweep efficiency and early breakthrough because of gravity override and viscous fingering effects (Dellinger et al. 1984; Alquriaishi and Shokir 2011). Moreover, high capital and operating costs are required to continuous CO_2 flooding in many field applications (Holt et al. 2009).

WAG injection as shown in Fig. 1.1 was proposed to improve sweep efficiency of CO_2 and mobility control, and the first field application was conducted in the

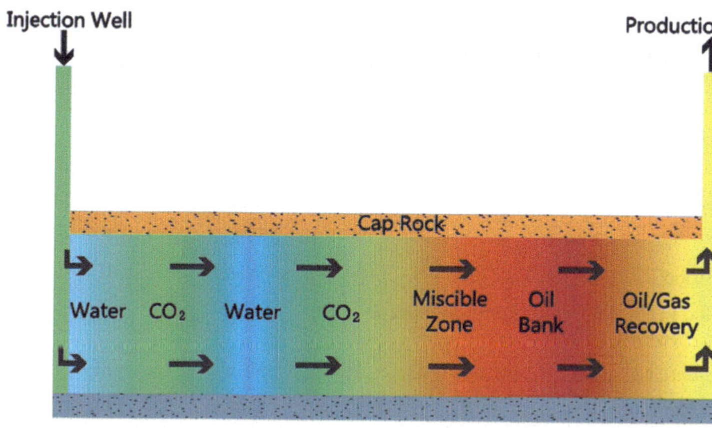

Fig. 1.1 Schematic diagram of water alternating gas (WAG) process using CO_2

North Pembina field in Alberta, Canada in 1957 (Christensen et al. 2001). WAG injection has a number of advantages over continuous CO_2 flooding. Firstly, injected water during CO_2 WAG can maintain the reservoir pressure above the MMP, and it makes the miscible reservoir condition easily. Secondly, WAG injection improves volumetric sweep efficiency compared to continuous CO_2 injection by controlling the mobility and stabilizing the fluid front. Also, CO_2 WAG may considerably reduce CO_2 consumption resulting in economizing on capital and operating costs. Finally, WAG injection is strongly recommended for carbon capture and storage (CCS) due to the improved residual and solubility trapping performances (Sifuentes et al. 2009). However, WAG injection scheme causes a complex saturation pattern because water and gas saturations consecutively change. Three-phase relative permeability is necessary to accurately describe the fluid flow during WAG. The majority of three-phase relative permeability models are summarized in Table 1.1 (Beygi et al. 2015).

1.2 Carbon Capture, Utilization, and Storage

Climate change concerns have caused global efforts to reduce the concentration of CO_2 in the atmosphere, so various methods are considered to reduce CO_2 emissions (Leung et al. 2014). First, improving energy efficiency and promoting energy conservation. It is related to reduce energy consumption. Second, increasing usage of low carbon fuels and renewable energy for residential and commercial demand, industry, and electricity generation. These methods are associated to energy supply and consumption. Lastly, carbon capture and storage (CCS) is considered a crucial strategy for reducing CO_2 emission directly. Each method for reducing CO_2 emission has advantages and limitations as shown in Table 1.2 (Leung et al. 2014).

Table 1.1 Three-phase relative permeability numerical models to describe fluid flow of oil, water, and gas in porous media (Beygi et al. 2015)

Model	Wettability	Phase applicability
Corey et al. (1956)	Water	3P
Wylie and Gardner (1958)	Water	Oil and gas
Naar and Wygal (1961)	Water	3P
Land (1968)	Water	3P
Stone I and Stone II (Stone 1970, 1973)	Water	3P
Modified Stone I (Hirasaki) and Modified Stone II (Dietrich and Bondor 1976)	Water	3P
Modified Stone I and Stone II (Aziz and Settari 1979)	Water	Oil
Parker and Lenhard (1987)	Water	Oil
Alemán and Slattery (1988)	Water	Oil
Baker (1987, 1988)	Water	Oil
Delshad and Pope (1989)	Water	Oil
Modified Stone I (Kokal and Maini 1990)	Water	Oil
Modified Stone I (Hustad and Holt 1992)	Water	Oil
Dria et al. (1993)	Water	4P
Robinson and Slattery (1994)	Water	Oil
Bradford et al. (1997)	FW	Water and oil
Jerauld (1997)	Int.	3P
Moulu et al. (1997, 1999)	All	3P
Goodyear and Townsley (Balbinski et al. 1999)	Water	Oil
DiCarlo et al. (2000)	Water + Int.	Oil
Modified Baker (Blunt 2000)	All	3P
Modified Baker (Fayers et al. 2000)	All	3P
Van Dijke et al. (2001)	All + FW	3P
Hustad and Hansen (1995), Hustad (2002), Hustad and Browning (2010)	All	3P
Shahverdi and Sohrabi (2012)	All	3P
Yuan and Pope (2012)	All	3P
UTKR3P (developed model) (2013)	All	3P

3P water/oil/gas; *4P* water/oil/gas/second hydrocarbon (CO_2/oil mixing at low temperature and reservoir pressures); *all* water-wet, mixed-wet, and oil-wet; *FW* Fractional wet; *Int.* Intermediate wet

Among the above methods, CCS is the only technology to reduce CO_2 significantly from the use of fossil fuels. It can reduce emissions not only from power generation, but also from industrial sectors (IEA 2015a). While other methods bring less or zero CO_2 emissions, CCS delivers negative emission. It becomes increasingly important for the carbon credit as temperature targets approach 2 °C and below after Paris

Table 1.2 Various methods for reducing CO_2 emissions (Leung et al. 2014)

Strategy	Application area/sector	Advantages	Limitations
Enhance energy efficiency and energy conservation	Applied mainly in commercial and industrial buildings	Energy saving from 10 to 20% easily achievable	May involve extensive capital investment for installation of energy saving device
Increase usage of clean fuels	Substitution of coal by natural gas for power generation	Natural gas emits 40–50% less CO_2 than coal due to its lower carbon content and higher combustion efficiency; cleaner exhaust gas (lower particulates and sulfur dioxide emissions)	Higher fuel cost for conventional natural gas. Comparable cost for shale gas
Adopt clean coal technologies	Integrated gasification combined cycle (IGCC), pressurized fluidized bed combustor (PFBC), etc., to replace conventional combustion	Allow the use of coal with lower emissions of air pollutants	Significant investment needed to roll out technologies widely
Use of renewable energy	Hydro, solar (thermal), wind power, and biofuels highly developed	Use of local natural resources; no or low greenhouse and toxic gas emissions	Applicability may depend on local resources availability and cost. Power from solar, wind, marine, etc., are intermittent and associated technologies are not mature; most current renewable energies are more costly than conventional energy

(continued)

Table 1.2 (continued)

Strategy	Application area/sector	Advantages	Limitations
Development of nuclear power	Nuclear fission is adopted mainly in U.S.A., France, Japan, Russia, and China Nuclear fusion is still in research and development phase	No air pollutant and greenhouse gas emissions	Usage is controversial; development of world's nuclear power is hindered due to the Fukushima Nuclear Accident in 2011, e.g., Germany will phase out all its nuclear power by 2022
Afforestation and reforestation	Applicable to all countries	Simple approach to create natural and sustainable CO_2 sinks	Restricts/prevents land use for other applications
Carbon capture and storage	Applicable to large CO_2 point emission sources	It can reduce vast amount of CO_2 with capture efficiency >80%	CCS full chain technologies not proven at full commercial scale

Agreement. CCS has been proven in many applications for more than 20 years through many experiences. There are 43 commercial large-scale global CCS facilities including 18 in operation, 5 in construction, and 20 in various stages of development as shown in Fig. 1.2 (Liu 2018).

The drawback of CCS in aquifer is the lack of economic feasibility because there is no profitability model. Coupled CO_2-EOR and CCS (CCS-EOR) could help each other for business and climate change mitigation because it can offer commercial

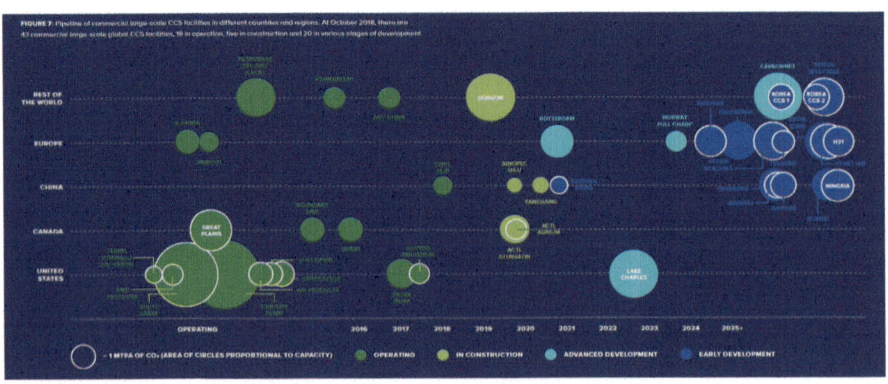

Fig. 1.2 Pipeline of commercial large-scale CCS facilities in different regions (area of circles proportional to capacity) (Liu 2018)

opportunities for oil producers and also ensures the permanent underground storage of large quantities of CO_2, so nearly 80 $MtCO_2$ is already being used for CCS-EOR each year (IEA 2015b, 2019). CCS-EOR is a promising short-to medium-term solution to alleviate the fast growth of CO_2 emissions (Metz et al. 2005; Ettehadtavakkol et al. 2014).

When only CCS is applied in aquifer, only two-phase (gas/water) system is necessary. In contrast, the four-phase (oil/water/gas/solid) system should be considered because CO_2 is injected into a reservoir and it changes oil composition causing asphaltene precipitation. Asphaltenes are the polar, polyaromatic, and high molecular weight hydrocarbon fractions of crude oil. They are generally insoluble in n-hexane or in n-pentane, but soluble in aromatic chemicals such as toluene or benzene (Srivastava and Huang 1997; Shedid and Abbas 2005; Frenier et al. 2010). When the oil in place is light to medium with small asphaltene contents, asphaltenes precipitate in reservoirs from oil due to changes of pressure, temperature, or oil composition during oil production (Gonzalez et al. 2014). Heavier crudes containing a larger amount of asphaltene have very few asphaltene precipitation problems because they can dissolve more asphaltene. Also, the maximum asphaltene precipitation occurs around saturation pressure. As pressure decreases before reaching saturation pressure, oil swelling causes the reduction of asphaltene solubility. After the reservoir oil releases solution gas below saturation pressure, oil becomes denser to cause asphaltene re-dissolution. Once asphaltenes have been precipitated from the oil, asphaltenes are deposited onto the rock surface. Causing pore plugging, porosity and permeability reduction, and wettability change, the deposition has been a serious problem during oil production in the field (Leontaritis and Mansoori 1988; Darabi et al. 2012). Change of oil composition causes asphaltene precipitation and deposition to occur during both primary production and miscible gas flooding with CO_2 (Srivastava et al. 1999; Jamaluddin et al. 2002; Sarma 2003; Dahaghi et al. 2008; Tahami et al. 2014). Asphaltene deposition on pore walls leads to wettability change and pore plugging, affecting not only EOR but also CCS simultaneously (Ju et al. 2010; Cho et al. 2019). Since it changes fluid flow, gas trapping mechanisms, and oil production, numerical modeling which reflects the mechanisms is necessary to predict CCS-EOR performances precisely. In this book, the processes for compositional fluid modeling and reservoir simulation in four-phase fluid system and three-phase hysteresis with asphaltene deposition effects will be described in detail.

References

Alemán MA, Slattery JC (1988) Estimation of three-phase relative permeabilities. Transp Porous Media 3(2):111–131. https://doi.org/10.1007/BF00820340

Alquriaishi A, Shokir EE-M (2011) Experimental investigation of miscible CO_2 flooding. Pet Sci Technol 29(19):2005–2016. https://doi.org/10.1080/10916461003662976

Aziz K, Settari A (1979) Petroleum reservoir simulation. Applied Science Publishers, London

Bahadori A (2018) Fundamentals of enhanced oil and gas recovery from conventional and unconventional reservoirs. Gulf Professional Publishing, Oxford, United Kingdom

Baker L (1987) Review of three-phase relative permeability literature: data and correlation methods. Tulsa, Oklahoma

Baker L (1988) Three-phase relative permeability correlations. In: Paper presented at the SPE enhanced oil recovery symposium, Tulsa, Oklahoma, 16–21 April 1988. https://doi.org/10.2118/17369-ms

Balbinski EF, Fishlock TP, Goodyear SG, Jones PIR (1999) Key characteristics of three-phase oil relative permeability formulations for improved oil recovery predictions. Petroleum Geosci 5(4):339–346. https://doi.org/10.1144/petgeo.5.4.339

Beygi M, Delshad M, Pudugramam V, Pope G, Wheeler M (2015) Novel three-phase compositional relative permeability and three-phase hysteresis models. SPE J 20(01):21–34. https://doi.org/10.2118/165324-PA

Blunt MJ (2000) An empirical model for three-phase relative permeability. SPE J 5(04):435–445. https://doi.org/10.2118/67950-PA

Bradford SA, Abriola LM, Leij FJ (1997) Wettability effects on two-and three-fluid relative permeabilities. J Contam Hydrol 28(1–2):171–191

Cho J, Kim T, Chang N, Lee K (2019) Effects of asphaltene deposition-derived formation damage on three-phase hysteretic models for prediction of coupled CO_2 enhanced oil recovery and storage performance. J Petrol Sci Eng 172:988–997. https://doi.org/10.1016/j.petrol.2018.09.006

Christensen J, Stenby E, Skauge A (2001) Review of WAG field experience. SPE Reserv Eval Eng 4(02):97–106. https://doi.org/10.2118/71203-PA

Corey AT, Rathjens CH, Henderson JH, Wyllie MRJ (1956) Three-phase relative permeability. J Petrol Technol 8(11):63–65. https://doi.org/10.2118/737-G

Dahaghi A, Gholami V, Moghadasi J, Abdi R (2008) Formation damage through asphaltene precipitation resulting from CO_2 gas injection in Iranian carbonate reservoirs. SPE Prod Oper 23(02):210–214. https://doi.org/10.2118/99631-PA

Darabi H, Sepehrnoori K, Kalaei M (2012) Modeling of wettability alteration due to asphaltene deposition in oil reservoirs. In: Paper presented at the SPE annual technical conference and exhibition, San Antonio, Texas, USA, 8–10 Oct 2012. https://doi.org/10.2118/159554-ms

Dellinger S, Patton J, Holbrook S (1984) CO_2 mobility control. SPE J 24(02):191–196. https://doi.org/10.2118/9808-PA

Delshad M, Pope GA (1989) Comparison of the three-phase oil relative permeability models. Transp Porous Media 4(1):59–83. https://doi.org/10.1007/BF00134742

DiCarlo DA, Akshay S, Blunt MJ (2000) Three-phase relative permeability of water-wet, oil-wet, and mixed-wet sandpacks. SPE J 5(01):82–91. https://doi.org/10.2118/60767-PA

Dietrich JK, Bondor PL (1976) Three-phase oil relative permeability models. In: Paper presented at the SPE annual fall technical conference and exhibition, New Orleans, Louisiana, 3–6 Oct 1976. https://doi.org/10.2118/6044-ms

Dria DE, Pope GA, Sepehrnoori K (1993) Three-phase gas/oil/brine relative permeabilities measured under CO_2 flooding conditions. SPE Reserv Eng 8(02):143–150. https://doi.org/10.2118/20184-PA

Elmabrouk SK, Bader HE, Mahmud WM (2017) An overview of power plant CCS and CO_2-EOR projects. In: Paper presented at the international conference on industrial engineering and operations management, Rabat, Morocco, 11–13 April 2017

Ettehadtavakkol A, Lake L, Bryant S (2014) CO_2-EOR and storage design optimization. Int J Greenhouse Gas Control 25:79–92. https://doi.org/10.1016/j.ijggc.2014.04.006

Fayers FJ, Foakes AP, Lin CY, Puckett DA (2000) An improved three phase flow model incorporating compositional variance. In: Paper presented at the SPE/DOE improved oil recovery symposium, Tulsa, Oklahoma, 3–5 April 2000. https://doi.org/10.2118/59313-ms

Frenier W, Ziauddin M, Venkatesan R (2010) Organic deposits in oil and gas production. Society of Petroleum Engineers, Richardson, Texas, USA

Gonzalez K, Barrufet M, Nasrabadi H (2014) Development of a compositional reservoir simulator including asphaltene precipitation from a thermodynamic consistent model. In: Paper presented at the SPE Latin America and Caribbean petroleum engineering conference, Maracaibo, Venezuela, 21 May 2014. https://doi.org/10.2118/169401-ms

Gozalpour F, Ren S, Tohidi B (2005) CO_2 EOR and storage in oil reservoir. Oil Gas Sci Technol Rev IFP 60:537–546. https://doi.org/10.2516/ogst:2005036

Han L, Gu Y (2015) Miscible CO_2 water-alternating-gas (CO_2-WAG) injection in a tight oil formation. In: Paper presented at the SPE annual technical conference and exhibition, Houston, Texas, USA, 28 Sept 2015. https://doi.org/10.2118/175108-ms

Holt T, Lindeberg E, Wessel-Berg D (2009) EOR and CO_2 disposal—economic and capacity potential in the North Sea. Energy Procedia 1(1):4159–4166. https://doi.org/10.1016/j.egypro.2009.02.225

Hustad ØS (2002) A coupled model for three-phase capillary pressure and relative permeability. SPE J 7(01):59–69. https://doi.org/10.2118/74705-PA

Hustad OS, Browning DJ (2010) A fully coupled three-phase model for capillary pressure and relative permeability for implicit compositional reservoir simulation. SPE J 15(04):1003–1019. https://doi.org/10.2118/125429-PA

Hustad OS, Hansen AG (1995) A consistent correlation for three phase relative permeabilities and phase pressures based on three sets of two phase data. In: Paper presented at the 8th European IOR symposium, Vienna, Austria, 15 May 1995. https://doi.org/10.3997/2214-4609.201406940

Hustad OS, Holt T (1992) Gravity stable displacement of oil by hydrocarbon gas after waterflooding. In: Paper presented at the SPE/DOE enhanced oil recovery symposium, Tulsa, Oklahoma, 22–24 April 1992. https://doi.org/10.2118/24116-ms

IEA (2015a) Carbon capture and storage: the solution for deep emissions reductions. IEA, Paris

IEA (2015b) Storing CO_2 through enhanced oil recovery: combining EOR with CO_2 storage (EOR+) for profit. IEA, Paris

IEA (2019) Transforming industry through CCUS. IEA, Paris

Jamaluddin A, Joshi N, Iwere F, Gurpinar O (2002) An investigation of asphaltene instability under nitrogen injection. In: Paper presented at the SPE international petroleum conference and exhibition in Mexico, Villahermosa, Mexico, 10–12 Feb 2002. https://doi.org/10.2118/74393-ms

Jerauld GR (1997) General three-phase relative permeability model for prudhoe bay. SPE Reserv Eng 12(04):255–263. https://doi.org/10.2118/36178-PA

Johns R, Dindoruk B, Orr FM (1993) Analytical theory of combined condensing/vaporizing gas drives. SPE Adv Technol Ser 1(02):7–16. https://doi.org/10.2118/24112-PA

Ju B, Qiu X, Qin J, Chen X, Fan T (2010) Asphaltene deposition and its effects on production performances in the development of oil field by CO_2 flooding: a numerical simulation assessment. In: Paper presented at the SPE EUROPEC/EAGE annual conference and exhibition, Barcelona, Spain, 14–17 June 2010. https://doi.org/10.2118/129804-ms

Kamali F, Hussain F, Cinar Y (2017) An experimental and numerical analysis of water-alternating-gas and simultaneous-water-and-gas displacements for carbon dioxide enhanced oil recovery and storage. SPE J 22(02):521–538. https://doi.org/10.2118/183633-PA

Kokal S, Maini BB (1990) An improved model for estimating three-phase oil-water-gas relative permeabilities from two-phase oil-water and oil-gas data. J Can Pet Technol 29(02):11. https://doi.org/10.2118/90-02-07

Land CS (1968) Calculation of imbibition relative permeability for two- and three-phase flow from rock properties. SPE J 8(02):149–156. https://doi.org/10.2118/1942-PA

Leontaritis K, Mansoori G (1988) Asphaltene deposition: a survey of field experiences and research approaches. J Petrol Sci Eng 1(3):229–239. https://doi.org/10.1016/0920-4105(88)90013-7

Leung DYC, Caramanna G, Maroto-Valer MM (2014) An overview of current status of carbon dioxide capture and storage technologies. Renew Sustain Energy Rev 39:426–443. https://doi.org/10.1016/j.rser.2014.07.093

Liu H (2018) The global status of CCS. Global CCS Institute, Australia

Makimura D, Kunieda M, Liang Y, Matsuoka T, Takahashi S, Okabe H (2013) Application of molecular simulations to CO_2-enhanced oil recovery: phase equilibria and interfacial phenomena. SPE J 18(02):319–330. https://doi.org/10.2118/163099-PA
Metz B, Davidson O, de Coninck H, Loos M, Meyer L (2005) IPCC special report on carbon dioxide capture and storage. IPCC, United States
Moulu JC, Vizika O, Egermann P, Kalaydjian F (1999) A new three-phase relative permeability model for various wettability conditions. In: Paper presented at the SPE annual technical conference and exhibition, Houston, Texas, 3–6 Oct 1999. https://doi.org/10.2118/56477-ms
Moulu JC, Vizika O, Kalaydjian F, Duquerroix JP (1997) A new model for three-phase relative permeabilities based on a fractal representation of the porous medium. In: Paper presented at the SPE annual technical conference and exhibition, San Antonio, Texas, 5–8 Oct 1997. https://doi.org/10.2118/38891-ms
Naar J, Wygal RJ (1961) Three-phase imbibition relative permeability. SPE J 1(04):254–258. https://doi.org/10.2118/90-PA
Parker JC, Lenhard RJ (1987) A model for hysteretic constitutive relations governing multiphase flow: 1. Saturation-pressure relations. Water Resources Res 23(12):2187–2196. https://doi.org/10.1029/wr023i012p02187
Robinson RL, Slattery JC (1994) Estimation of three-phase relative permeabilities. Transp Porous Media 16(3):263–287. https://doi.org/10.1007/BF00617151
Sarma H (2003) Can we ignore asphaltene in a gas injection project for light-oils? In: Paper presented at the SPE international improved oil recovery conference in Asia Pacific, Kuala Lumpur, Malaysia, 20–21 Oct 2003. https://doi.org/10.2118/84877-ms
Shahverdi H, Sohrabi M (2012) Three-phase relative permeability and hystresis model for simulation of water alternating gas (WAG) injection. In: Paper presented at the SPE improved oil recovery symposium, Tulsa, Oklahoma, USA, 14–18 April 2012. https://doi.org/10.2118/152218-ms
Shedid S, Abbas E (2005) An experimental approach of the reversibility of asphaltene deposition under dynamic flow conditions. In: Paper presented at the SPE Middle East oil and gas show and conference, Kingdom of Bahrain, 12–15 Mar 2005. https://doi.org/10.2118/92159-ms
Sifuentes W, Giddins M, Blunt M (2009) Modeling CO_2 storage in aquifers: assessing the key contributors to uncertainty. In: Paper presented at the offshore Europe, Aberdeen, UK, 8–11 Sept 2009. https://doi.org/10.2118/123582-ms
Srivastava R, Huang S (1997) Asphaltene deposition during CO_2 flooding: a laboratory assessment. In: Paper presented at the SPE production operations symposium, Oklahoma City, Oklahoma, 9–11 Mar 1997. https://doi.org/10.2118/37468-ms
Srivastava R, Huang S, Dong M (1999) Asphaltene deposition during CO_2 flooding. SPE Prod Facil 14(04):235–245. https://doi.org/10.2118/59092-PA
Stalkup F (1987) Displacement behavior of the condensing/vaporizing gas drive process. In: Paper presented at the SPE annual technical conference and exhibition, Dallas, Texas, 27–30 Sept 1987. https://doi.org/10.2118/16715-ms
Stone HL (1970) Probability model for estimating three-phase relative permeability. J Petrol Technol 22(02):214–218. https://doi.org/10.2118/2116-PA
Stone HL (1973) Estimation of three-phase relative permeability and residual oil data. J Can Pet Technol 12(04):10. https://doi.org/10.2118/73-04-06
Tahami S, Dabir B, Asghari K, Shahvaranfard A (2014) Modeling of asphaltene deposition during miscible CO_2 flooding. Pet Sci Technol 32(18):2183–2194. https://doi.org/10.1080/10916466.2010.504937
Van Dijke MIJ, McDougall SR, Sorbie KS (2001) Three-phase capillary pressure and relative permeability relationships in mixed-wet systems. Transp Porous Media 44(1):1–32. https://doi.org/10.1023/A:1010773606657

Wylie M, Gardner G (1958) The generalized Kozeny-Carman equation: its application to problems of multiphase flow in porous media. World Oil 146:210–227

Yuan C, Pope G (2012) A new method to model relative permeability in compositional simulators to avoid discontinuous changes caused by phase-identification problems. SPE J 17(04):1221–1230. https://doi.org/10.2118/142093-PA

Chapter 2
Major Mechanisms

Abstract CCS-EOR has several features being dissimilar to conventional CCS process, which are more complicated to understand and describe in numerical methods. When the conventional CCS is conducted in aquifer, only two-phase system (water and gas) and solubility trapping mechanism are considered. However, four-phase system (water, gas, oil, and solid) should be applied in CCS-EOR because CO_2 and water are injected into oil reservoirs, and they cause solid precipitation and deposition of asphaltene that affect CCS and EOR performances and mechanisms. In this chapter, the important features to model CCS-EOR process such as methodologies of hydrocarbon fluid modeling, minimum miscibility pressure (MMP) calculations, asphaltene modeling, formation damages, and four-phase hysteresis model are presented. Firstly, various equation of state (EOS) models are introduced, and the definition and modeling method of MMP follows. Then, the solubility, thermodynamic micellization, solid, and association EOS models for modeling asphaltene precipitation are introduced with some characteristics and limitations. Precipitated asphaltenes may deposit in reservoirs causing formation damages such as permeability and porosity reduction and wettability alteration that affect CCS-EOR performance by changing hysteresis and solubility mechanisms. Therefore, solubility, three-phase permeability, and hysteresis models are described.

Keywords Equation of state (EOS) · Minimum miscibility pressure (MMP) · Miscibility · Asphaltene deposition · Solid model · Three-phase hysteresis

2.1 Hydrocarbon System

2.1.1 Van der Waals Equation of State

An equation of state (EOS) is a thermodynamic equation to describe the state and relationship between pressure, volume, and temperature (PVT). An appropriate description of PVT is necessary to determine the phase behavior and predict the properties of fluids in petroleum reservoirs. The simplest form of EOS is the ideal gas equation as below:

K. S. Lee et al., *CO_2 Storage Coupled with Enhanced Oil Recovery*,
https://doi.org/10.1007/978-3-030-41901-1_2

$$p = \frac{RT}{V} \tag{2.1}$$

where p and T are absolute pressure and temperature, R is the universal gas constant, and V is the molar volume. Two assumptions were made in the ideal gas EOS. Firstly, the volume of gas molecules and distance between the molecules were ignored. Secondly, the attractive and repulsive forces between molecules were not considered. Since the ideal gas equation has a limitation that it only can be applied to the near atmospheric pressure condition, a number of EOS suitable for describing the phase behavior of substances at extended ranges of pressures and temperatures have been proposed.

Van der Waals EOS (1873) tried to overcome the assumptions of ideal gas EOS. Considering the volume of gas molecules because it occupies a significant fraction of the volume at higher pressures, he could eliminate the first assumption of ideal gas EOS. It can be expressed in the following equation by using parameter b which reflects the volume of molecules or repulsion parameter.

$$p = \frac{RT}{V - b} \tag{2.2}$$

The second assumption of ideal gas EOS was removed by considering attractive forces between molecules in Van der Waals EOS. The final form of EOS is shown as below:

$$p = \frac{RT}{V - b} - \frac{a}{V^2} \tag{2.3}$$

The first term of right-hand side indicates repulsive pressure term, and the last term represents the attractive power. Since the gas phase volume is greater than that of molecules at low pressure, the parameter b can be negligible. Also, the term of $\frac{a}{V^2}$ does not need to be considered because attractive force is not dominant in low pressure condition. Therefore, the cubic EOS equation can be reduced to ideal gas EOS at low pressure. Parameters a and b for pure components can be determined by the characteristic that the value of first and second differential equation in P-V plane becomes zero as indicated in Fig. 2.1. It also can be expressed in the equations as follows:

$$\left[\frac{\partial p}{\partial V}\right]_{T_c, p_c} = 0 \tag{2.4}$$

$$\left[\frac{\partial^2 p}{\partial V^2}\right]_{T_c, p_c} = 0 \tag{2.5}$$

Above equations are usually transformed to a general form for calculating a and b:

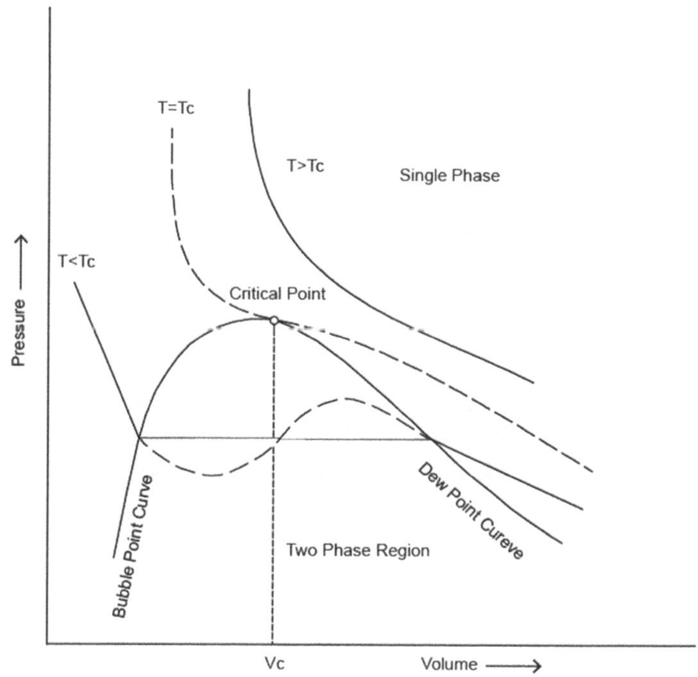

Fig. 2.1 Pressure and volume (PV) diagram for pure fluid

$$a = \Omega_a \frac{R^2 T_c^2}{p_c} \tag{2.6}$$

$$b = \Omega_b \frac{R T_c}{p_c} \tag{2.7}$$

where Ω_a and Ω_b are 0.421875 and 0.125 for Van der Waals EOS. Equation 2.3 can be expressed in a cubic form in terms of volume V as below:

$$V^3 - \left(b + \frac{RT}{p}\right)V^2 + \left(\frac{a}{p}\right)V - \left(\frac{ab}{p}\right) = 0 \tag{2.8}$$

The equation is usually used to estimate the volume of liquid and gas phases. For a pure substance at a constant temperature below critical point, a typical volumetric behavior is in indicated P-V plot of Fig. 2.2. When Eq. 2.8 is solved for V, it has three real roots solutions for each specific pressure p, and each solution of volume refers to point A, D, and F in Fig. 2.2. Between A and F along the line, liquid and gas phases exist simultaneously, so volume D does not have physical meaning. Point A, the smallest solution for Van der Waals cubic EOS corresponds to the volume of saturated liquid. In contrast, the largest root as shown by point F signifies the volume of saturated gas because gas volume is much greater than liquid at the specific pressure

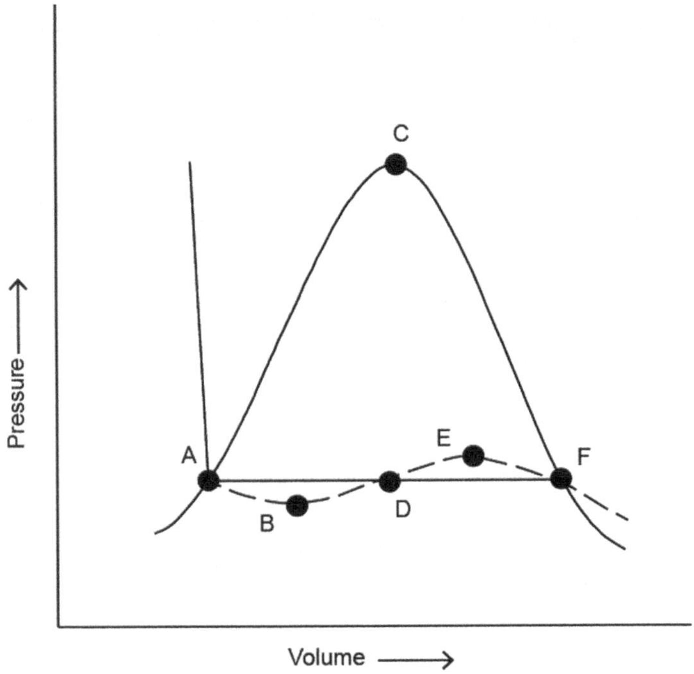

Fig. 2.2 Pressure and volume (PV) diagram of a pure component

and temperature. Volumes of saturated liquid and gas are getting closer as system temperature becomes similar to critical temperature, and they are finally the same as point C in Fig. 2.2 when the temperature reaches the critical one. The above equation, which is expressed in volume *V*, is usually reproduced in terms of the compressibility factor *Z* as below:

$$Z^3 - (1 + B)Z^2 + AZ - AB = 0 \tag{2.9}$$

where

$$A = \frac{ap}{R^2T^2} \tag{2.10}$$

$$B = \frac{bp}{RT} \tag{2.11}$$

As similar to the solution of volume cubic equation, the largest and smallest roots are the compressibilities of vapor and liquid phases among the solutions of Eq. 2.9. Finally, liquid and gas densities can be calculated by using the compressibility factors in the following equation:

$$\rho = \frac{pM}{ZRT} \tag{2.12}$$

where M is the molecular weight of the component.

2.1.2 Redlich and Kwong Equation of State

The attractive term, $\frac{a}{V^2}$ in Van der Waals EOS was improved by Redlich and Kwong (1949), and it is called RK EOS. The term is changed to temperature-dependent term as follows:

$$p = \frac{RT}{V - b} - \frac{a}{V(V + b)\sqrt{T}} \tag{2.13}$$

where

$$a = \Omega_a \frac{R^2 T_c^{2.5}}{p_c} \tag{2.14}$$

$$b = \Omega_b \frac{RT_c}{p_c} \tag{2.15}$$

and Ω_a and Ω_b are 0.42747 and 0.08664, respectively. At high pressures, the volumes of all gases approach a limiting value which is practically independent of the temperature and is close to 26% of critical volume, V_c, so b can be also expressed as 0.26 V_c. Therefore, the following equation is also satisfied.

$$p_c V_c = 0.333 RT_c \tag{2.16}$$

and it indicates that the critical compressibility factor Z_c for all components is 0.333 in RK EOS.

Equation 2.13 is usually represented in terms of Z factor as below:

$$Z^3 - Z^2 + \left(A - B - B^2\right)Z - AB = 0 \tag{2.17}$$

where

$$A = \frac{ap}{R^2 T^{2.5}} \tag{2.18}$$

$$B = \frac{bp}{RT} \tag{2.19}$$

As in Van der Waals EOS, the largest and smallest real roots are the compressibility factors of gas and liquid phases.

Van der Waals EOS shows the phase behavior of only a single substance, but RK EOS extended it to hydrocarbon mixture. The mixture parameters in liquid and gas phases are shown as follows:

$$a_m = \left(\sum_{i=1}^{n} x_i \sqrt{a_i}\right)^2 \text{ or } \left(\sum_{i=1}^{n} y_i \sqrt{a_i}\right)^2 \tag{2.20}$$

$$b_m = \sum_{i=1}^{n} x_i b_i \text{ or } \sum_{i=1}^{n} y_i b_i \tag{2.21}$$

where

n number of components in mixture
i component number of mixture
x_i mole fraction of component i in liquid phase
y_i mole fraction of component i in gas phase.

Equations 2.20 and 2.21 can be substituted into Eqs. 2.18 and 2.19 to calculate the mixture parameters A and B, and compressibility factor Z of mixture is estimated by Eq. 2.17. The mixture density is calculated by Eq. 2.12 with apparent molecular weight as below:

$$M_m = \sum_{i=1}^{n} x_i M_i \tag{2.22}$$

2.1.3 Soave-Redlich-Kwong Equation of State

Though RK EOS has good accuracy for volumetric and thermal properties of pure component and mixtures, its VLE (Vapor liquid equilibrium) prediction of multi-component gives poor results. Soave (1972) tried to improve the vapor pressure calculation of pure substances. The original RK EOS has been modified by replacing the term $\frac{1}{\sqrt{T}}$ in Eq. 2.13, to give:

$$p = \frac{RT}{V - b} - \frac{a\alpha}{V(V + b)} \tag{2.23}$$

where α is a temperature dependent and dimensionless term, and it is defined as follows:

$$\alpha = \left[1 + m\left(1 - \sqrt{T_r}\right)\right]^2 \tag{2.24}$$

α becomes 1 when system temperature reaches critical temperature because reduced temperature T_r is the ratio of system and critical temperature. The parameter m is as below:

$$m = 0.480 + 1.574\omega - 0.176\omega^2 \tag{2.25}$$

where ω is an acentric factor of the substance, and Pitzer et al. (1955) proposed it to predict the vapor pressure of pure substance more accurately. The acentric factor is a comprehensive representation of the characteristics of the boiling point, molecular weight, and polarity of pure component. The longer molecule with the deviation from the perfect sphere and non-polarity has the greater acentric factor value. The parameter a and b in Eq. 2.23 can be expressed as that of Van der Waals, such as Eqs. 2.6 and 2.7, but Ω_a and Ω_b are 0.42747 and 0.08664 for SRK EOS.

SRK EOS is usually represented in terms of Z factor as Eq. 2.17 in RK EOS, but parameters A and B can be given by:

$$A = \frac{a\alpha p}{R^2T^2} \tag{2.26}$$

$$B = \frac{bp}{RT} \tag{2.27}$$

EOS equations are applied to describe the phase equilibrium that is gained by the condition of chemical equilibrium. For a two-phase system (liquid and gas), the chemical potential of each pure component in liquid phase should be the same as that in gas phase, and chemical potential is usually expressed in terms of fugacity f which can be defined as below:

$$f = p\ \exp\left[\int_0^p \left(\frac{Z-1}{p}\right)dp\right] \tag{2.28}$$

The pressure p on the right-hand side can be moved to the left, and the ratio of the fugacity and pressure, dimensionless term, is called fugacity coefficient, φ. When the Eq. 2.23 is applied into the above equation and solved for the fugacity coefficient, the general form for pure component is as below:

$$\ln\ \varphi = \ln\left(\frac{f}{p}\right) = Z - 1 - \ln(Z - B) - \frac{A}{B}\ln\left(1 + \frac{B}{Z}\right) \tag{2.29}$$

The concept of fugacity is usually used for multi-component equilibrium involving solid, liquid, and gas phases. It is practically useful as an engineering tool for predicting how much the components are in each phase in multi-component mixtures at various temperatures and pressures without the actual lab test. If the fugacity of the component i in the first phase is different from that in the second phase, the phase transfer of i occurs. The phase with the lower component fugacity will receive the component from the phase with the higher one until the fugacity in each phase becomes the same as follows:

$$f_i^V = f_i^L, \quad 1 \le i \le n \tag{2.30}$$

where f_i^V and f_i^L are the fugacities in vapor and liquid phases, and n is the number of components in the system.

Reid et al. (1977) defined the fugacity coefficient of a component in a hydrocarbon mixture. When the SRK EOS is applied into it, the final form for the fugacity coefficient of component i is as follows:

$$\ln \varphi_i = \frac{B_i}{B}(Z-1) - \ln(Z-B) + \frac{A}{B}\left(\frac{B_i}{B} - \frac{2}{A}\sum_{j=1}^{n} y_i A_{ij}\right)\ln\left(1 + \frac{B}{Z}\right) \tag{2.31}$$

where y_i is the mole fraction of vapor phase, and it can be transferred to x_i for that of liquid phase. The parameters A, B, and A_{ij} are expressed as following equations for the mixture by using mixing rules:

$$A = \sum_{i=1}^{n}\sum_{j=1}^{n} y_i y_j A_{ij} \tag{2.32}$$

$$B = \sum_{i=1}^{n} y_i B_i \tag{2.33}$$

$$A_{ij} = \left(1 - k_{ij}\right)\sqrt{A_i A_j} \tag{2.34}$$

where k_{ij} is the binary interaction parameter.

2.1.4 Peng–Robinson Equation of State

Peng and Robinson (1976) evaluated and improved the SRK EOS to predict the behavior of hydrocarbon more accurately. They proposed a two-constant equation that can estimate the liquid density and fluid properties near a critical point. Peng–Robinson equation (PR EOS) is given by:

$$p = \frac{RT}{V-b} - \frac{a\alpha}{V(V+b) + b(V-b)} \tag{2.35}$$

or in terms of Z factor:

$$Z^3 - (B-1)Z^2 + \left(A - 3B^2 - 2B\right)Z - \left(AB - B^2 - B^3\right) = 0 \tag{2.36}$$

where Z_c is 0.3074. The parameters A and B are given by Eqs. 2.26 and 2.27 for pure components. Moreover, a, b, and α in Eq. 2.35 also can be expressed by equations that are used in SRK EOS, but Ω_a and Ω_b are 0.45724 and 0.07780 for PR EOS. In PR EOS, the parameter m in the equation for α is given in the following equation:

$$m = 0.37464 + 1.54226\omega - 0.26992\omega^2 \tag{2.37}$$

Robinson and Peng (1978) proposed a modified m for heavier components that acentric factor is greater than 0.49 as below:

$$m = 0.379642 + 1.48503\omega - 0.1644\omega^2 + 0.016667\omega^3 \tag{2.38}$$

The Z factor equation can be solved by an analytical or a trial and error method (Edmister and Lee 1984). The fugacity expression of component i in a hydrocarbon mixture for PR EOS is given by:

$$\ln \varphi_i = \frac{B_i}{B}(Z - 1) - \ln(Z - B) + \frac{A}{2\sqrt{2}B}\left(\frac{B_i}{B} - \frac{2}{A}\sum_{j=1}^{n} y_i A_{ij}\right)\ln\left[\frac{Z + \left(1 + \sqrt{2}\right)B}{Z - \left(1 - \sqrt{2}\right)B}\right] \tag{2.39}$$

The above equation can be used to determine the fugacity coefficient in liquid phase by replacing the composition of gas phase y_i with that of the liquid phase x_i.

The PR EOS and SRK EOS are the two most widely used cubic equations of state. They deliver the accuracy for vapor liquid equilibrium and volumetric predictions when volume translation is used (Whitson and Brule 2000). Peneloux et al. (1982) introduced a volume correction parameter, c_i, to improve the volumetric predictions of the SRK EOS. This volume shift process does not affect equilibrium calculations for pure components and mixtures. A volume correction term in multi-component system is expressed as below:

$$V_L^{\text{corr}} = V_L^{\text{EOS}} - \sum_{i=1}^{n} x_i c_i \tag{2.40}$$

$$V_V^{\text{corr}} = V_V^{\text{EOS}} - \sum_{i=1}^{n} y_i c_i \tag{2.41}$$

where V_L^{corr} and V_V^{corr} are molar volumes of liquid and vapor phases corrected by volume translation, and V_L^{EOS} and V_V^{EOS} are liquid and vapor molar volumes that are calculated by EOS. Moreover, x_i and y_i are mole fraction of component i in liquid and vapor phases, respectively. Whitson and Brule (2000) improved Jhaveri

Table 2.1 Volume shift parameters for the PR and SRK EOS (Ahmed 2010)

Compound	PR EOS	SRK EOS
N_2	−0.1927	−0.0079
CO_2	−0.0817	0.0833
H_2S	−0.1288	0.0466
C_1	−0.1595	0.0234
C_2	−0.1134	0.0605
C_3	−0.0863	0.0825
i-C_4	−0.0844	0.0830
n-C_4	−0.0675	0.0975
i-C_5	−0.0608	0.1022
n-C_5	−0.0390	0.1209
n-C_6	−0.0080	0.1467
n-C_7	0.0033	0.1554
n-C_8	0.0314	0.1794
n-C_9	0.0408	0.1868
n-C_{10}	0.0655	0.2080

and Youngren's work (1988) and suggested the volume shift parameter, c_i, for some pure components as shown in Table 2.1.

Hydrocarbon mixture may exist as a single or two phases in reservoirs. EOS can be applied to two-phase flash calculation which can estimate the amounts and compositions of hydrocarbon liquid and gas phases at a given pressure, temperature, and overall composition. The procedure for flash calculation is as below (Ahmed 2010):

1. Assume the initial K values for each component in the mixture at the given pressure and temperature by using Wilson's equation (1968) as follows:

$$K_i^A = \frac{p_{ci}}{p} \exp\left[5.37(1 + \omega_i)\left(1 - \frac{T_{ci}}{T}\right)\right]$$

2. Perform flash calculation to calculate the number of moles and composition of each component in liquid and gas phases by using overall composition and the assumed K values.
3. Determine the fugacity coefficient for each component in the liquid and gas phases with its composition.
4. Calculate the new equilibrium ratios as below:

$$K_i^C = \frac{\varphi_i^L}{\varphi_i^V}$$

5. Compare the ratio of assumed and calculated K values if it is within a tolerance

$$\sum_{i=1}^{n}\left(\frac{K_i^C}{K_i^A}-1\right)^2 \leq \varepsilon$$

where ε is a preset error tolerance which is usually 0.0001, and n is the number of components.

Steps 1 to 5 will be repeated by using the result of step 4 as an initial K value, and when the result of step 5 in the above processes is accepted, it will be the solution.

2.2 Minimum Miscibility Pressure

2.2.1 Definition of Miscibility

CO_2 injection, as an efficient EOS process, has been applied for a long time to recover the remaining oil by reduction of the interfacial tension and oil viscosity, extracting light component, and maintaining the reservoir pressure (Jha 1986; Rojas and Ali 1988; Alvarado and Manrique 2010; Cao and Gu 2013; Rommerskirchen et al. 2016). The CO_2 flooding can be performed with a miscible or an immiscible displacement process (Wang and Gu 2011). Though main mechanisms of immiscible and miscible flood are similar, the effects, such as oil swelling and viscosity reduction, are weaker in immiscible flood than that in miscible one because of the limited miscibility between CO_2 and oil (Cho et al. 2015). The criterion to verify that the reservoir reaches miscible or immiscible condition during CO_2 flooding is minimum miscibility pressure (MMP), which is not only for CO_2 injection but also for any solvent gas to displace residual oil (Ghorbani et al. 2014). Miscible gas injection can be only applied when the reservoir pressure is greater than MMP, whereas immiscible flood will be conducted under MMP condition. Since MMP is a very important factor in gas flooding process to predict the performance of the process, an accurate MMP estimation is always essential for reservoir engineers (Golkari and Riazi 2017).

While CO_2 flooding is conducted, miscibility can be achieved by two methods, first contact and multiple contact miscible processes. Condition of first or multiple contact miscibility at a specified pressure depends on the injected gas and oil composition and temperature (Green and Willhite 1998; Gu et al. 2013). Figure 2.3 shows the ternary diagram of phase behavior of reservoir oil and injected gas at a given pressure and temperature, and the inner circle area indicates two-phase zone. If the straight line of injected gas and reservoir oil passes the two-phase zone, it indicates that the first contact miscibility does not occur. When the injection pressure is sufficiently high, the injected CO_2 can be miscible with the oil at the first contact, and the pressure is called as the first contact miscibility pressure. Therefore, the first contact miscibility is reached when reservoir pressure is high enough to make two-phase zone smaller until

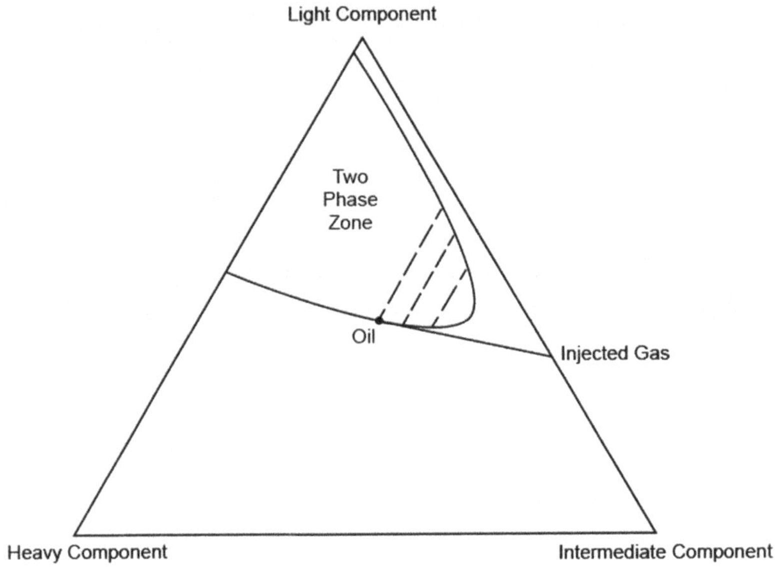

Fig. 2.3 Ternary diagram of phase behavior of reservoir oil and injected gas at a given pressure and temperature for first contact miscibility

the line oil does not pass the two-phase zone. Also, if injected gas is heavy enough as indicated in Fig. 2.3, such as liquefied petroleum gas (LPG) which consists of propane and butane, first contact miscible condition can be achieved (Stalkup 1983).

It is difficult to achieve the first contact miscibility between reservoir oil and CO_2 due to limitation of injection pressure. Multiple contact miscibility process usually occurred in real field. The miscibility condition is achieved by mass transfer between the injected gas and the reservoir oil with multiple contact, and the physical properties of two phases become gradually close to each other. When CO_2 is injected into reservoirs, CO_2 is consistently contacted with fresh oil following flow direction at the fore-end of injected fluid, and they are miscible by the vaporizing gas drive process. Figure 2.4 shows the mechanism of vaporizing gas drive miscibility. The composition of reservoir oil A lies on the extension of the limiting tie line from the plait point P. Injected gas and reservoir oil are not miscible at the first contact, and overall composition becomes M_1 that consists of liquid L_1 and gas G_1 in equilibrium condition. Equilibrium gas G_1 contacts fresh reservoir oil after the first contact at the fore-end of gas stream, and the composition of G_1 changes to G_2 by vaporizing intermediate hydrocarbons from the oil. Intermediate molecular weight hydrocarbons are extracted from the fresh reservoir oil multiple times, so injected gas properties are continuously changed until reaching the plait point P that indicates miscible condition.

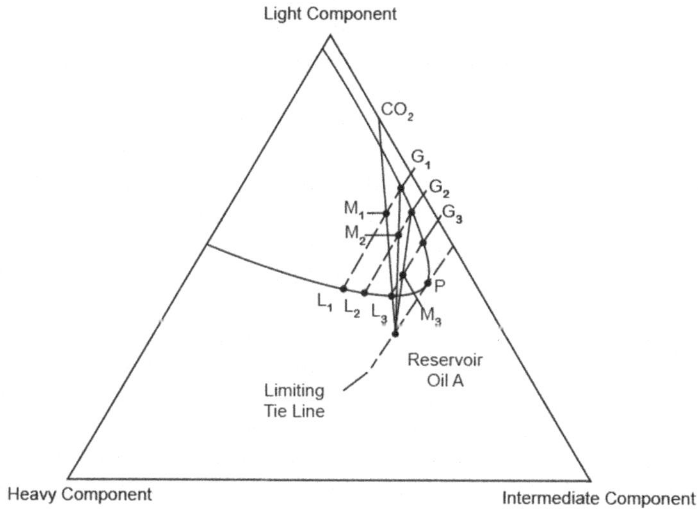

Fig. 2.4 Ternary diagram of phase behavior of reservoir oil and injected gas at a given pressure and temperature for vaporizing gas drive mechanism

Contrastively, reservoir oil is continuously contacted with fresh CO_2 at the back-end of injected CO_2, and it causes the miscible state by the condensing gas drive process. In this process, dynamic miscibility occurs with the transfer of intermediate molecular weight hydrocarbon from the injected gas to the reservoir oil. When the gas A contacts the oil at the first time, they are not miscible because the line passes two-phase zone in Fig. 2.5. Similar to the vaporizing gas drive mechanism, the mixture overall composition after the first contact becomes M_1 with gas G_1 and oil L_1. The residual oil L_1 at the back-end of injected gas stream may contact fresh gas continuously, and the composition keeps changing until it reaches the plait point P. The schematic diagram of multiple contact miscibility process is indicated in Fig. 2.6. The miscibility developments are usually reached through the vaporizing, condensing, or combined vaporizing and condensing gas drive. The mechanisms of the multiple contact miscibility depend on the temperature and fluid compositions, so they are varying for different cases. Therefore, it can be incorrect if we simply conclude that specific mechanism always occur in the reservoir.

2.2.2 *Slim-Tube Experiments*

A number of methods including slim tube, multiple contract mixing cell, vanishing interfacial tension, and rising bubble experiments have traditionally been used to estimate MMP between crude oil and injected gas (Yellig and Metcalfe 1980; Christiansen and Haines 1987; Rao 1997; Ghorbani et al. 2014). Among them, slim tube

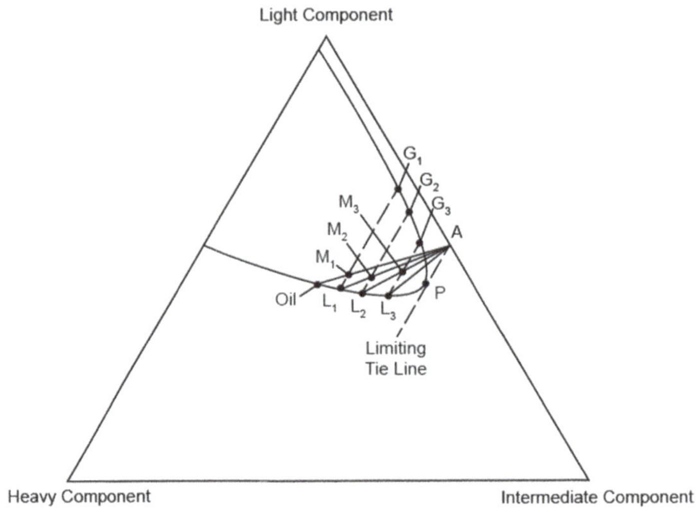

Fig. 2.5 Ternary diagram of phase behavior of reservoir oil and injected gas at a given pressure and temperature for condensing gas drive mechanism

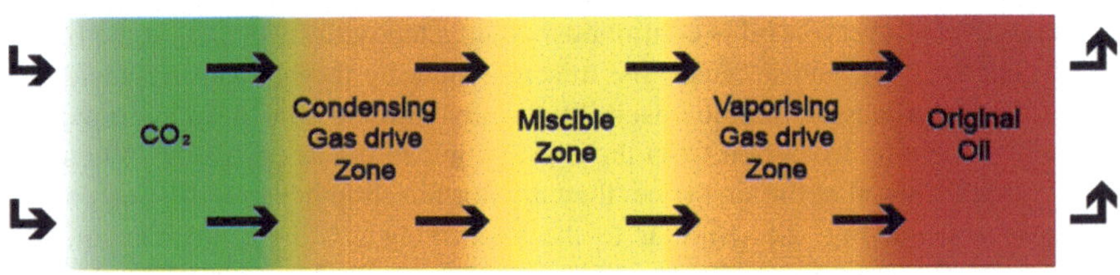

Fig. 2.6 The schematic diagram of multiple contact miscibility process

test has been the most standard and widely used for decades in experimental ways (Amao et al. 2012). The slim tube is a cylindrical tube which is packed with uniform sand or glass beads. It is initially saturated with the reservoir oil above bubble point pressure to set the system condition as a single phase (Randall and Bennion 1989; Wu and Batycky 1990). The gas is injected at different system pressures and temperatures through the slim tube to model the oil and gas fluid flow. The test is performed in 1-D fluid flow condition to eliminate the viscous fingering and entrance effect. All slim tube tests at various pressures are terminated after 1.0–2.0 PV of gas is injected. Most of them were finished after 1.2 PV of gas injection, and the oil recovery factors are measured at each pressure condition. In early studies, the oil recovery was the only criterion to determine the MMPs from the slim-tube tests. The miscibility was achieved if the oil recovery factor was over 80% at CO_2 breakthrough or 94% at the end of a slim-tube test, and the MMP was estimated at the system pressure (Holm and Josendal 1974; Randall and Bennion 1988). Although different oil recovery factors can be measured for the criterion to determine the MMP, 90% or a higher oil recovery

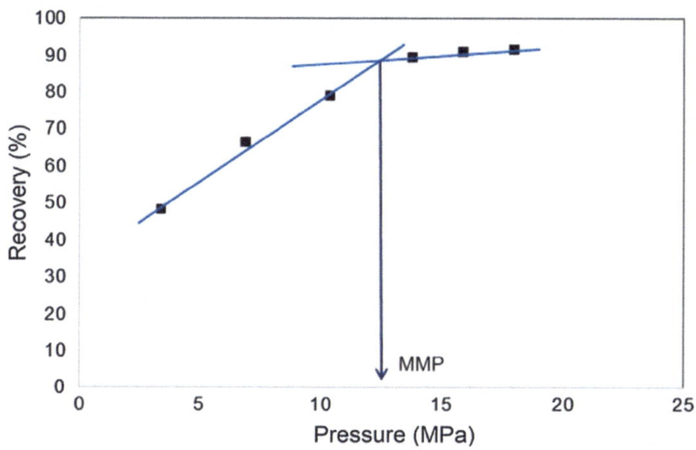

Fig. 2.7 Schematic for determination of MMP using slim tube tests (Le Van and Chon 2017)

at 1.2 PV of injected gas was commonly accepted to be an indicator (Metcalfe 1982). The definition of MMP as the distinct point of maximum curvature when cumulative oil recovery at 1.2 PV gas injection is plotted with pressure is still widely used as indicated in Fig. 2.7 (Le Van and Chon 2017). However, slim-tube experiments are expensive and time-consuming that they take weeks to perform one set of miscibility experiments (Ekundayo and Ghedan 2013). Therefore, some qualified numerical methods to calculate the MMP have been developed to enhance efficiency, but the results should be verified with experimental data for reliability.

2.2.3 *Multiple Mixing Cell Method*

Because of the disadvantages of experimental methods, computational methods, such as analytical calculations, slim-tube compositional simulation, and multi-contact mixing cell models, have been developed for MMP calculation (Monroe et al. 1990; Johns et al. 1993; Neau et al. 1996; Wang and Orr 1997; Jaubert et al. 1998; Zhao et al. 2006; Orr 2007). Multiple mixing cell method by using cubic EOS is one of the most prevalent ways to estimate MMP (Ahmadi and Johns 2011). The method is simple, practical, and robust with any number of components. It is simple because recovery factor or transfer amounts of fluid are not necessary to calculate MMP and only flash calculation is needed. According to Ahmadi and Johns (2011) showing the results of the method for some fluid samples, the calculated MMPs by multiple mixing cell method are well matched to that of slim-tube experiments. The illustration of repeated contact for the method is indicated in Fig. 2.8, and the method is applied with the procedures given below:

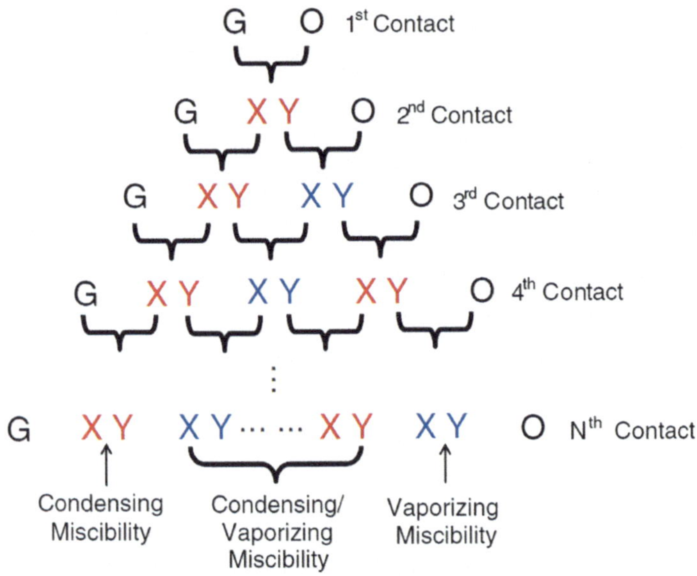

Fig. 2.8 Schematic for determination of MMP using slim tube tests (Schematic multiple contact in the multiple mixing cell method. *G* injected gas composition, *O* oil composition, *Y* equilibrium gas composition, *X* equilibrium oil composition) (Ahmadi and Johns 2011)

1. Specify the reservoir initial pressure and temperature, usually 500 psia for the pressure.
2. Perform flash calculation with injection gas and reservoir oil to obtain two new equilibrium compositions of liquid and vapor after first contact.
3. Calculate the tie-line length for the pressure by using the equation below:

$$TL = \sqrt{\sum\nolimits_{i=1}^{N_c} (x_i - y_i)^2}$$

4. Continue Steps 2 and 3 with additional contacts by mixing the next cells until all key tie lines converge to a tolerance.
5. Increase the pressure and repeat the Step 2 through 4.
6. Calculate the length of each key tie line for each pressure and store the minimum tie line length (*TL*).
7. Perform a multiple parameter regression of *TL* and draw the graph of *TL* as a function of pressures to determine n in $TL^n = ap + b$ (power-law extrapolation) where a and b are the slope and constant, respectively.
8. Determine the MMP when the power-law extrapolation gives zero of *TL* within the desired accuracy of 20 psia at the last three pressures.

2.3 Asphaltene Model

2.3.1 Introduction to Asphaltene

Molecular structures of asphaltene are shown in Fig. 2.9. They are the polar, polyaromatic, and high molecular weight hydrocarbon fraction of crude oil that are generally insoluble in paraffins like n-pentane or heptane, but soluble in aromatic chemicals, such as toluene or benzene. The asphaltene dissolution can lose its stabilizing equilibrium due to the change of reservoir pressure, temperature, and oil composition, so that the asphaltene components may get separated from the crude and precipitate (Jamaluddin et al. 2001; Oskui et al. 2009). Precipitated asphaltenes flow as a solid particle suspended in the liquid phase. Once asphaltene precipitation occurs, deposition may ensue in the formation, tubing, pipelines, and surface facilities.

Because the reservoir is usually isothermal condition unless thermal EOS is applied, the consideration to investigate the effect of temperature change on asphaltene precipitation is not necessary during CCS-EOR process. In the view of the

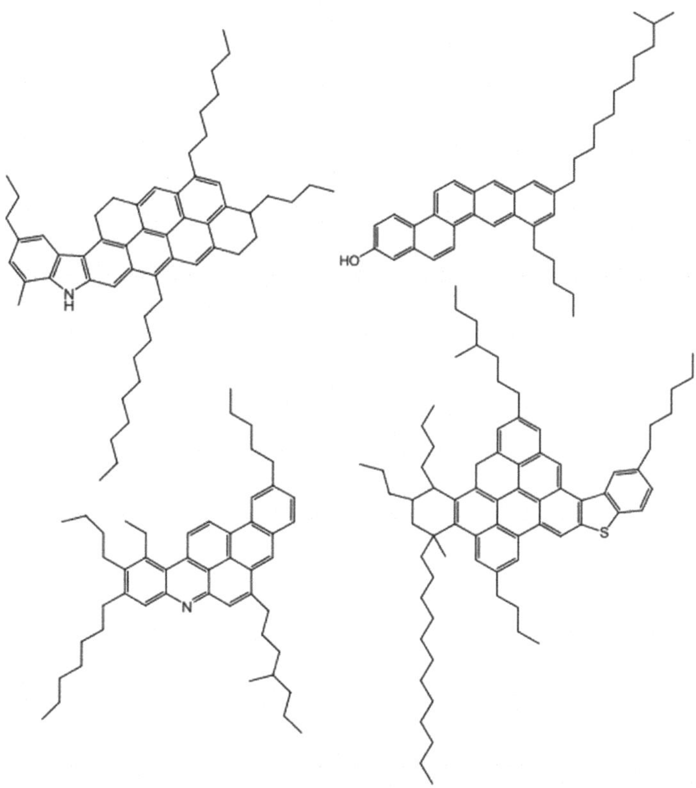

Fig. 2.9 Asphaltene molecular structures (Mullins 2005)

pressure change, asphaltene starts to precipitate at the onset point pressure during oil production. As pressure decreases before getting saturation pressure, the oil density gradually declines, resulting in the reduction of asphaltene solubility. Therefore, the maximum asphaltene precipitation happens around the saturation pressure because of the lowest oil density. After the reservoir oil releases solution gas below saturation pressure, oil becomes denser causing asphaltene re-dissolution as indicated in Fig. 2.10, and the asphaltene precipitation envelop showing the deposition stages is shown in Fig. 2.11. Precipitation occurs only within the region between the asphaltene onset pressure upper boundary and the lower boundary.

The injection of light components into the reservoir increases the instability of asphaltene dissolution in oil phase causing precipitation around the reservoir and in the wellbore (Darabi et al. 2014). Therefore, CO_2 injection affects the onset pressure and the amount of asphaltene precipitation, which have been proved by various experimental and numerical ways (Srivastava et al. 1999; Gonzalez et al. 2008; Adyani et al. 2011; Mahmoudvand et al. 2019).

The precipitated asphaltenes can flocculate and deposit anywhere in the reservoir (Fig. 2.12). The deposited asphaltenes lead to formation damages, such as permeability and porosity reductions and wettability alteration of the rock surface that eventually affect reservoir EOR and CCS performances and mechanisms (Cho et al. 2019). Therefore, it should be considered to evaluate the performances and assets.

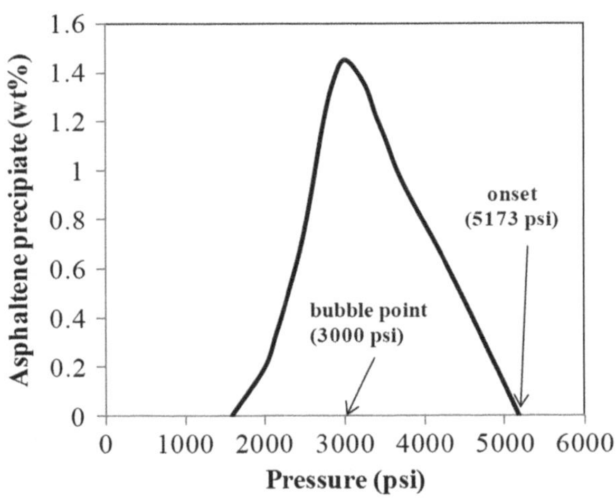

Fig. 2.10 Asphaltene precipitation curve as a function of pressure (Darabi et al. 2012)

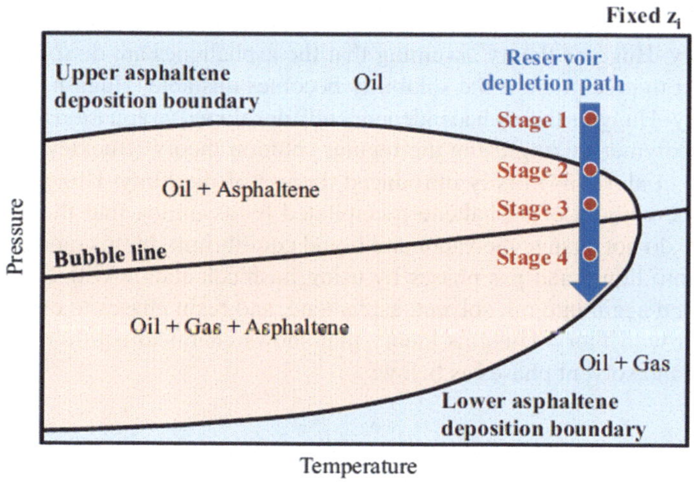

Fig. 2.11 p-T asphaltene precipitation envelope showing the deposition stages in a reservoir (Gonzalez et al. 2017)

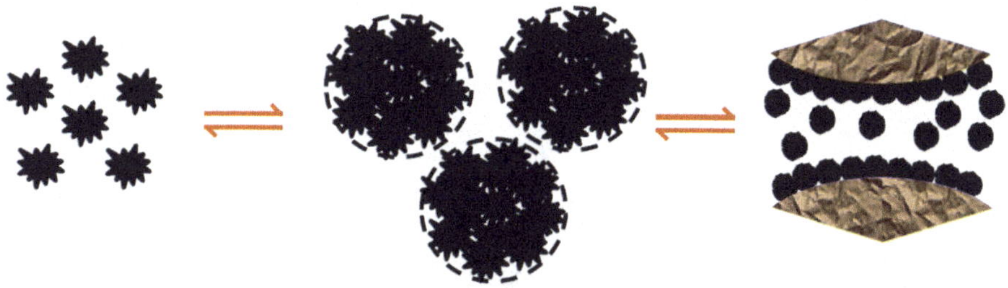

Fig. 2.12 Schematic of asphaltene precipitation, flocculation, and deposition processes (Darabi et al. 2012)

2.3.2 Asphaltene Modeling

Estimating the amount of asphaltene precipitation in experimental ways is very challenging because it is time-consuming and expensive. Therefore, various thermodynamic models have been developed to predict the onset and offset point pressure and the amount of asphaltene precipitation. The models are categorized into four different models, such as solubility, thermodynamic micellization, solid, and association EOS models (Akbarzadeh et al. 2007; Ahmadi et al. 2015).

Solubility models use the concept of solubility parameter and assume that crude consists of with and without asphaltene oils. Asphaltene solubility parameters are calculated by the interaction parameters between the asphaltene and the remaining oil components (Subramanian et al. 2016). Solubility models are based on the simplified Flory–Huggins theory assuming that the asphaltenes are dissolved in crude oil and start to precipitate if the solubility becomes unstable (Huggins 1941; Flory 1942). Flory–Huggins theory had independently developed an equation for Gibbs free energy of polymer mixing using the regular solution theory (Buckley et al. 2007). Hirschberg et al. (1984) firstly introduced the use of the Flory–Huggins theory to calculate the amount of asphaltene precipitated by assuming that the precipitated asphaltenes do not change the vapor and liquid equilibrium. In this model, fluids are classified into liquid and gas phases by using flash calculation with EOS, and liquid is divided again into oil, solvent, asphaltene, and resin phases to estimate phase equilibrium with Flory–Huggins theory that shows chemical equilibrium between asphaltene and solvent phases as below:

$$\mu_i^{\text{asp}} = \mu_i^{\text{sol}} \tag{2.42}$$

where chemical potential of i component, μ_i, is calculated by the following equation:

$$\frac{\mu_i - \mu_{i,\text{ref}}}{RT} = \ln(\varphi_i) + 1 - \frac{V_i}{V_m} + \chi_i \tag{2.43}$$

$\mu_{i,\text{ref}}$ is the chemical potential at the reference state of pure liquid, and the volume fraction of component i, φ_i, the molar volume of the mixture, V_m, and the interaction parameter, χ_i, are as follows:

$$\varphi_i = \frac{x_i V_i}{V_m} \tag{2.44}$$

$$V_m = \sum x_i V_i \tag{2.45}$$

$$\chi_i = \frac{V_i}{RT}(\delta_m - \delta_i)^2 \tag{2.46}$$

where x_i is the mole fraction of component i and the solubility parameter of mixture, δ_m, is given by:

$$\delta_m = \sum \varphi_i \delta_i \tag{2.47}$$

The solubility parameter δ is calculated from Hildebrand and Scott's definition (1950):

$$\delta^2 = \frac{\Delta u^v}{v} \tag{2.48}$$

where v represents the molar volume and Δu^v is the internal energy change when one mole of liquid is vaporized to gas phase which is gained on the basis of SRK

EOS. Finally, the maximum volume fraction of asphaltene soluble in crude oil is obtained from the equation:

$$\ln(\varphi_{\text{asp}})\left(1 - \frac{v_{\text{asp}}}{v_m}\right)\frac{v_{\text{asp}}}{RT}(\delta_{\text{asp}} - \delta_m)^2_{\max} \tag{2.49}$$

There are many kinds of solubility models which are simple to use, but they are not predictive (Nor-azlan and Adewumi 1993; Cimio et al. 1995; Kraiwattanawong et al. 2007). Thermodynamic micellization models are developed by Victorov and Firoozabadi (1996) and Pan and Firoozabadi (1997) that assume asphaltene molecules have a micelle core with the absorbed resin molecules on the surface of this core. The structure and concentration of the micelle are calculated by minimizing the Gibbs free energy of micellization. Though the models have good agreement of the size of asphaltene micelle with the experimental data, the amount of asphaltene precipitation cannot be presented.

The new association EOSs have been developed to model complex fluid phase behavior including asphaltene component. The major models are Cubic Plus Association (CPA), Statistical Association Fluid Theory (SAFT), Perturberd Chain SAFT (PC-SAFT) EOSs. The CPA EOS was developed by Kontogeorgis et al. (1996) and it describes the associating fluids by dividing fluids into two parts, a physical and an associating terms. CPA EOS was recently used by Li and Firoozabadi (2010) and Shirani et al. (2012) to model asphaltene precipitation. SAFT model, where molecules are modeled as chains of bonded spherical segments, was developed by Chapman et al. (1988, 1989, 1990) based on extensions and simplifications of Wertheim's theory for associating fluids (Wertheim 1984a, b, 1986a, b). In SAFT, the molecules are modeled in form of chains that consist of bonded spherical segments as shown in Fig. 2.13. SAFT EOS was modified by Gross and Sadowski (2001) to PC SAFT, where chain molecules were used as reference fluid instead of spherical molecules. Many researchers have modeled asphaltene by using PC-SAFT EOS and studied the effects of injection gases on asphaltene phase behavior (Gonzalez et al. 2005, 2007; Schou Pedersen and Hadsbjerg 2007; Gonzalez Rodriguez 2008; Vargas et al. 2009; Panuganti et al. 2013; Mohebbinia et al. 2014; Zuniga-Hinojosa et al. 2014;

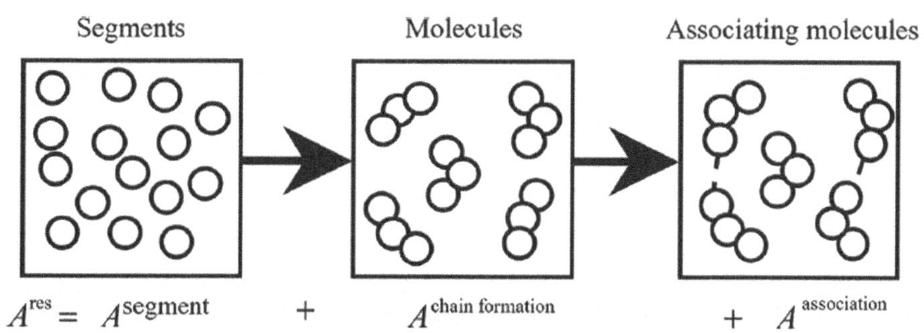

Fig. 2.13 Molecular models for SAFT EOS (Subramanian et al. 2016)

Sabeti et al. 2015). Though the estimation of asphaltene precipitation behavior is more accurate with association EOSs due to consideration for the interaction of the molecules, the models are not fully developed for fluid characterization (Darabi et al. 2014).

Gupta (1986) firstly proposed a solid model to estimate the amount of asphaltene precipitation by using the cubic EOS. It assumes the precipitated asphaltene as a pure solid component. The model has been widely used in the reservoir simulators until now because it is fast and shows good agreement with experimental data even though it has a limitation of inability to indicate the effect of polar or association components (Chung 1992; Nghiem et al. 1993; Nghiem and Coombe 1997; Kohse et al. 2000; Nghiem et al. 2000). In this model, the heaviest component of oil phase is divided into two components: precipitating and non-precipitating components (Kohse and Nghiem 2004). These components have the same critical properties and acentric factors, but different interaction coefficient parameters with the lighter components. While fugacities of oil and gas phases containing non-precipitating component are calculated by PR EOS, the fugacity of precipitated asphaltenes in solid phase S_1 is estimated by:

$$\ln\ f_{S_1} = \ln\ f_{S_1}^* + \frac{v_{S_1}(p - p^*)}{RT} \tag{2.50}$$

where

v_{S_1} solid phase molar volume
p^* onset pressure of asphaltene precipitation at T
$f_{S_1}^*$ fugacity of solid phase at p^* and T.

The equilibrium condition is achieved when the fugacities of all components in oil, gas, and solid phases are equal as follows:

$$\ln\ f_{i,o} = \ln\ f_{i,g},\quad i = 1, \ldots, n_c \tag{2.51}$$

$$\ln\ f_{n_c,o} = \ln\ f_{S_1} \tag{2.52}$$

and S_1 is reversible because of the thermodynamic equilibrium conditions.

Precipitated asphaltene splits into two solids S_1 and S_2 by applying reversible chemical reaction (Kohse and Nghiem 2004). Colloidal asphaltenes are flocculated to S_2 depending on the reaction rate as below:

$$r = \frac{dC_{S_2}}{dt} = K_{12}C_{S_1,o} - K_{21}C_{S_2,o} \tag{2.53}$$

where

$C_{S_1,o}$ concentration of the colloidal asphaltenes in oil phase
$C_{S_2,o}$ concentration of the flocculated asphaltenes in oil phase
K_{12} forward rate for the formation of S_2 from S_1

K_{21} backward rate for the formation of S_1 from S_2.

Some flocculated asphaltenes flow as suspended particles in oil phase, and the others deposit on the rock surface in reservoirs. Asphaltene deposition is calculated by Wang and Civan (2001)'s model which describes surface deposition, re-entrainment, and pore throat plugging with the following equation:

$$\frac{V_{S_2^d}^{n+1} - V_{S_2^d}^{n}}{\Delta t} = \alpha C_{S_2^f}^{n+1} \phi^{n+1} - \beta V_{S_2^d}^{n+1} \left(v_o^n - v_{cr,o} \right) + \gamma u_o^n C_{S_2^f}^{n+1} \tag{2.54}$$

where

$V_{S_2^d}$ volume of deposited solid S_2 per grid block
$C_{S_2^f}$ volumetric concentration of flowing solid S_2 per volume of oil
ϕ porosity
v_o oil phase interstitial velocity
$v_{cr,o}$ critical oil phase interstitial velocity
u_o oil phase Darcy velocity
n time step.

The first term of the right-hand side indicates surface deposition, and α is the surface deposition rate coefficient which depends on the rock type. The second term is re-entrainment of deposited asphaltenes into oil, and re-entrainment rate coefficient, β, is set to zero when v_o is smaller than $v_{cr,o}$. γ is the pore throat plugging coefficient calculated from the following equation and comes to be zero when the pore throat diameter is larger than critical value.

$$\gamma = \gamma_i \left(1 + \sigma V_{S_2^d} \right) \tag{2.55}$$

where

γ_i instantaneous pore throat plugging rate coefficient
σ snowball-effect deposition constant.

2.3.3 *Permeability and Porosity Reduction Model*

Deposited asphaltenes in reservoirs may firstly decrease reservoir permeability and porosity as one of formation damages. The reduced porosity and permeability after asphaltene deposition are calculated by using deposited asphaltene volume and a power-law function as follows (Reis and Acock 1994):

$$\phi = \phi_0 - V_{S_2^d} \tag{2.56}$$

$$\frac{k_0}{k} = \left(\frac{\phi_0}{\phi}\right)^n = F_{r,o} \quad (2.57)$$

where

ϕ_0 initial porosity
k_0 initial permeability
$F_{r,o}$ oil resistance factor.

The power law exponent n is usually between 3 and 7, and 5 is usually applied as an average value (Reis and Acock 1994).

2.4 Reservoir Model

2.4.1 Three-Phase Relative Permeability Models

Numerical models of three-phase relative permeability are necessary to describe fluids flow in porous media during CCS-EOR process because water and gas are injected in oil reservoirs. Many researchers have developed numerical models for relative permeability over the decades as indicated in Table 1.1 in Chap. 1. Since the measurement is very difficult and incomplete, three-phase relative permeability is conventionally estimated by use of measured two-phase relative permeability (Hajilary and Shahmohammdi 2018). Three-phase residual saturations depend on various factors such as wettability, IFT, fluid composition, saturation history, porosity, etc., and three-phase residual saturation models are indicated in Table 2.2 (Salathiel 1973; Suzanne et al. 2003; Pentland et al. 2010: Iglauer et al. 2012). These models use the saturation-averaged interpolation method instead of capillary, statistical, and network models (Beygi et al. 2015). Stone's first and second models (1970, 1973), traditional numerical models, have limited applicability in oil wet condition (Fayers and Matthews 1984; Delshad and Pope 1989; Hustad and Holt 1992; Pejic and Maini 2003; Spiteri et al. 2008; Ahmadloo et al. 2009). Although the limitation of Stone's models, these are modified by some researchers and mainly applied in commercial reservoir program combined with three-phase hysteresis (Dietrich and Bondor 1976; Aziz and Settari 1979; Kokal and Maini 1990; Hustad and Holt 1992; Thomas and Coats 1992; CMG 2018).

Stone's first model, which is modified by Aziz and Settari (1979), is usually implemented in commercial reservoir simulators. In this model, oil relative permeability is estimated by using that in oil-water and gas-liquid two-phase relative permeability tables, k_{row} and k_{rog}, and total oil relative permeability is calculated by:

$$k_{ro} = \frac{S_o^* k_{row} k_{rog}}{k_{rocw}\left(1 - S_w^*\right)\left(1 - S_g^*\right)} \quad (2.58)$$

Table 2.2 Three-phase residual saturation correlation models (Beygi et al. 2015)

Model	Rock wettability	Phase applicability	Residual-saturation correlation
Frayers and Matthews (1984)	Water	Oil	$S_{or}^{3P} = \chi S_{orw} + (1-\chi)S_{org}$, where $\chi = 1 - \frac{S_g}{1-S_{wc}-S_{org}}$
Fayers (1989)	All	Oil	$S_{or}^{3P} = S_{orw} - \alpha_w S_{gt}, 0 \leq \alpha_w \leq 1$
Fayers et al. (2000)	Mixed	Oil	$S_{or}^{3P} = \psi_r S_{org} + (1-\psi_r)S_{orw} + S_g \frac{S_w - S_{wc}}{\beta_F S_{orw} - \psi S_{gt}}$ where $\psi = \frac{S_g}{1-S_{wc}-S_o}$
Yuan and Pope (2012)	All	Water/Oil/Gas	$S_{jr}^{3P} = \min\left[S_j, S_{jr}^{2P}(1 - b_j S_k S_l)\right]$, where $j =$ 1, 2 and 3 and $j \neq k \neq l$

where k_{rocw} is the two-phase oil relative permeability at connate water saturation. Moreover, the saturation functions are gained from the equations as below:

$$S_o^* = \frac{S_o - S_{om}}{1 - S_{wc} - S_{om}} \tag{2.59}$$

$$S_w^* = \frac{S_w - S_{wc}}{1 - S_{wc} - S_{om}} \tag{2.60}$$

$$S_g^* = \frac{S_g}{1 - S_{wc} - S_{om}} \tag{2.61}$$

and the minimum oil saturation, S_{om}, is estimated from the linear function of gas saturation by using residual oil saturation in oil-water and gas-liquid two-phase relative permeability tables, S_{orw} and S_{org} (Fayers and Matthews 1984):

$$S_{om} = \alpha S_{orw} + (1 - \alpha) S_{org} \tag{2.62}$$

$$\alpha = 1 - \frac{S_g}{1 - S_{wc} - S_{org}} \tag{2.63}$$

where S_{wc} is connate water saturation.

2.4.2 Relative Permeability Estimation

Wettability of rock surface can be altered by various reasons including formation damages by asphaltene deposition. It controls multiphase fluid flow in CO_2 EOR or CCS-EOR processes (Treiber and Owen 1972; Morrow 1990; Buckley et al. 1997; Escrochi et al. 2008; Al-Aulaqi et al. 2011; Aziz 2011; Uetani 2014). Wettability alteration is represented by change in relative permeability (Buckley 1996; Al-Maamari and Buckley 2003; Darabi et al. 2014). In this book, modified relative permeability with contact angle and modified Brooks-Corey model are applied to investigate relative permeability change by asphaltene deposition as follows (Adibhatla et al. 2005; Lake et al. 2014):

$$k_{r,w} = k^*_{r,w}\left(\frac{S_w - S_{r,w}}{1 - S_{r,w} - S_{r,nw}}\right)^{e_w} \tag{2.64}$$

$$k^*_{r,w} = 1 + \left[k^{0*}_{r,w} + \frac{\cos\theta - \cos\theta^\circ}{\cos(\pi - \theta^0) - \cos\theta^\circ}\left(k^{0*}_{r,nw} - k^{0*}_{r,w}\right) - 1\right]\left(\frac{1 + T^0_w N^0_{T,w}}{1 + T_w N_{T,w}}\right) \tag{2.65}$$

$$e_w = 1 + \left[e^0_w + \frac{\cos\theta - \cos\theta^\circ}{\cos(\pi - \theta^0) - \cos\theta^\circ}\left(e^0_{nw} - e^0_w\right) - 1\right]\left(\frac{1 + T^0_w N^0_{T,w}}{1 + T_w N_{T,w}}\right) \tag{2.66}$$

where

k^*_r endpoint of relative permeability
S_r residual saturation
e_w exponential parameter
T_w trapping parameter
N_T trapping number.

Subscripts o, w, and nw signify initial condition, wetting, and non-wetting phases, respectively.

2.4.3 Residual Trapping with a Three-Phase Hysteresis Model

Two-phase hysteresis models are relatively simple and have been widely applied to describe the phenomenon, and most of them are based on the trapping model by Land (1968) and Carlson (1981). When this model is applied, the gas phase (non-wetting phase) hysteresis follows the theory of Land (trapping function) and Carlson (two-phase hysteresis model). Two-phase hysteresis models have a limit to be applied for CCS-EOR.

Hysteresis models for CCS based on two-phase (water/gas) system model are assumed to have reversibility of the relative permeability curves when the second cycle of drainage process occurs after imbibition. However, it is not validated with experiments for WAG process because a number of cycles of water and gas injection are conducted during WAG (Fatemi and Sohrabi 2012; Mahzari and Sohrabi 2016). Land (1968) and Carlson (1981)'s two-phase hysteresis models have been widely used. However, they are not appropriate to three-phase (water/gas/oil) system because the assumption of reversibility of the drainage and imbibition scanning curves is not adequate (Spiteri and Juanes 2006). Three-phase hysteresis model (WAG hysteresis) by Larsen and Skauge (1998), which is widely used and the only model available in commercial reservoir simulator, is applied in this book. The model has presented better match to experimental results than two-phase model (Shahrokhi et al. 2014; Chen et al. 2017). The drainage curves with gas injection follow the previous imbibition curves with water injection in two-phase hysteresis models (Fig. 2.14a). In contrast, gas relative permeabilities of following drainage processes are lower than that of previous imbibition indicating irreversibility, so gas has its own scanning curve at each drainage and imbibition process (Fig. 2.14b). All gas relative permeability scanning curves that is non-wetting phase in drainage process during increasing gas saturation are calculated by the equations from Larsen and Skauge (1998):

$$\left[k_{rg}^{\text{drain}}\left(S_g, S_w^I, S_g^{\text{start}}\right)\right]_n = \left\{\left[k_{rg}^{\text{input}}(S_g) - k_{rg}^{\text{input}}\left(S_g^{\text{start}}\right)\right] * \left(\frac{S_{wi}}{S_w^I}\right)^{\alpha}\right\}_n + \left[k_{rg}^{\text{imb}}\left(S_g^{\text{start}}\right)\right]_{n-1} \tag{2.67}$$

and those of decreasing gas saturation (non-wetting phase in imbibition process) are estimated from:

$$\left[k_{rg}^{\text{imb}}(S_g)\right]_n = \left[k_{rg}^{\text{drain}}\left(S_{gf}^{\text{trans}}\right)\right]_n \tag{2.68}$$

$$\left(S_{gf}^{\text{trans}}\right)_n = \left(S_{gf}\right)_n + \left(S_g^{\text{end}}\right)_{n-1} \tag{2.69}$$

$$\left(S_{gf}\right)_n = \frac{1}{2}\left[\left(S_g - S_{gr}\right) + \sqrt{\left(S_g - S_{gr}\right)^2 + \frac{4}{C^{\text{trans}}}\left(S_g - S_{gr}\right)}\right]_n \tag{2.70}$$

where

k_{rg}^{drain}	gas relative permeability of drainage processes
S_w^I	water saturation at the start of drainage process
S_g^{start}	gas saturation at the start of drainage process (end of previous imbibition)
$k_{\text{rg}}^{\text{input}}$	input gas relative permeability
S_{wi}	initial water saturation
k_{rg}^{imb}	gas relative permeability of imbibition process
α	reduction exponent

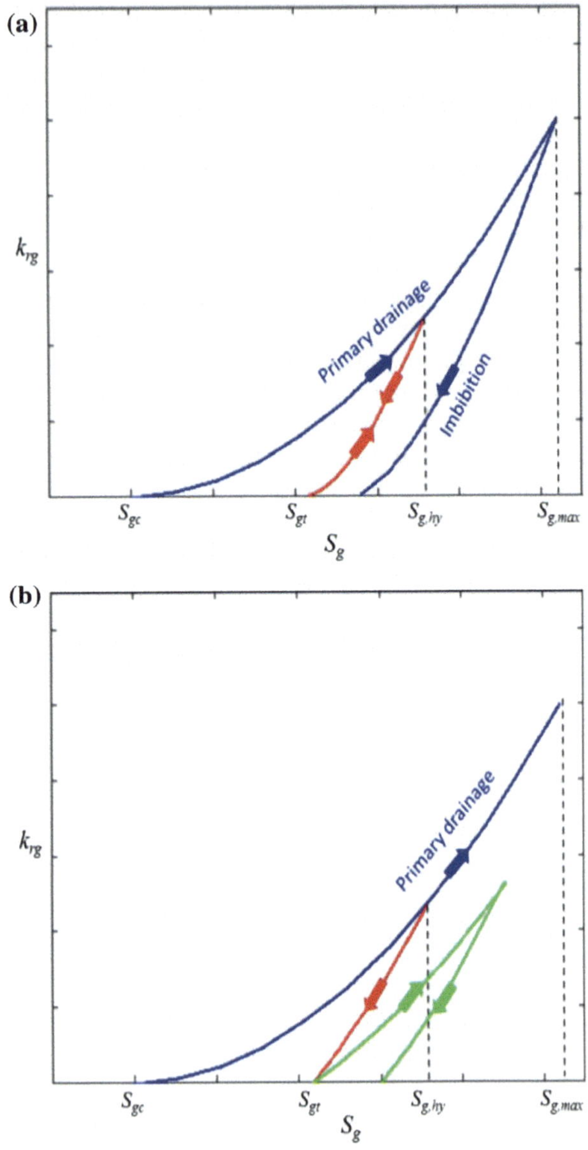

Fig. 2.14 Schematic of **a** two-phase and **b** three-phase hysteresis models with gas relative permeability (Chen et al. 2017)

S_{gf}^{trans} transformed free gas saturation
S_g^{end} endpoint gas saturation of imbibition.

The trapped gas saturation S_{gr}, Land's parameter (1968) C, and transformed Land's parameter C^{trans} at each step are given as:

$$S_{gr} = S_{gc} + \frac{S_{gm} - S_{gc}}{1 + C\left(S_{gm} - S_{gc}\right)} \tag{2.71}$$

$$C = \frac{1}{Sgc_{gr,\max}\frac{1}{Sgc_{g,\max}}} \tag{2.72}$$

$$C^{\text{trans}} = \frac{1}{\left(S_{gr}\right)_n - \left(S_g^{\text{end}}\right)_{n-1}} - \frac{1}{\left(S_{gi}\right)_n - \left(S_g^{\text{end}}\right)_{n-1}} \tag{2.73}$$

where

S_{gm} maximum gas saturation reached by drainage
S_{gc} critical gas saturation in input relative permeability table
$S_{g,\max}$ maximum gas saturation associated with the imbibition.

The maximum trapped gas saturation $S_{gr,\max}$ is calculated by use of porosity with an empirical correlation (Holtz 2002) as below, and the result of entire non-wetting phase hysteresis is indicated in Fig. 2.15.

$$S_{gr,\max} \tag{2.74}$$

To determine oil relative permeability in three-phase system by using two-phase data of water-oil and gas-oil, the Stone's first model, which is modified by Aziz and Settari (1979), is applied. The minimum residual oil saturation, S_{om}, is modified with trapped CO_2 saturation as follows:

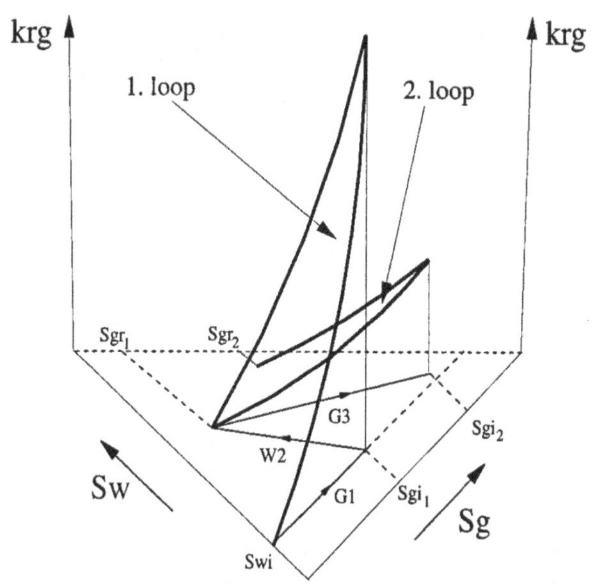

Fig. 2.15 Hysteresis loops of non-wetting phase (gas) during three-phase fluid flow (Larsen and Skauge 1998)

$$S_{om}\left(S_g - S_{gf}\right)_{om,mod} \tag{2.75}$$

where S_g and S_{gf} are total and movable gas saturations. Moreover, a is input parameter having values of 0.68 and 0.48 for strongly and slightly water-wet conditions, respectively (Mahzari and Sohrabi 2017).

2.4.4 Solubility Trapping Model

CO_2 is trapped in the reservoir by not only residual trapping but solubility trapping mechanism due to contact with reservoir oil and water. The mole fraction of CO_2 in each phase is determined from thermodynamic equilibrium condition as follows:

$$f_{i,o} = f_{i,g} = f_{i,w} \text{ with } i = 1, \ldots, n_c \tag{2.76}$$

where

$f_{i,o}$ fugacity of component i in oil phase
$f_{i,g}$ fugacity of component i in gas phase
$f_{i,w}$ fugacity of component i in water phase
n_c number of components in gas phase.

The fugacities of component i in gas and oil phases are calculated from Peng–Robinson EOS (1976, 1978). The fugacity of gas components soluble in water phase, in contrast, is attained by Henrys law (Li and Nghiem 1986) as below:

$$f_{iw} = y_{iw} H_i \text{ with } i = 1, \ldots, n_c \tag{2.77}$$

where

H_i Henry's constant of component i
y_{iw} mole fraction of component i in water phase.

Henry's constant is a function of pressure, temperature, and salinity. Generally, they all should be considered for CCS process in a thick aquifer due to the difference of temperature and salinity between the top and bottom of the aquifer. However, the influences of salinity and temperature change are ignored in CCS-EOR with LPG in the reservoir because it is assumed that injected water has zero salinity and constant temperature being same as the reservoir condition without change. Therefore, Henry's constant can be calculated by the following equations with isothermal condition (Saul and Wagner 1987; Harvey 1996):

$$\ln H_i = \ln H_i^s + \frac{1}{RT}\int_{p_{H_2O}^s}^{p} \bar{v}_i dp \tag{2.78}$$

$$\ln H_i^s = \ln p_{H_2O}^s - A\left(T_{r,H_2O}\right)^{-1} + B\left(1 - T_{r,H_2O}\right)^{0.355}\left(T_{r,H_2O}\right)^{-1} + C\,\exp\left(1 - T_{r,H_2O}\right)\left(T_{r,H_2O}\right)^{-0.41} \quad (2.79)$$

where

H_i^s	Henry's constant for component i at the saturation pressure of H_2O, MPa
$p_{H_2O}^s$	saturation pressure of H_2O at temperature T, MPa
T_{r,H_2O}	reduced temperature of H_2O
A	−9.4234 for CO_2
B	4.0087 for CO_2
C	10.3199 for CO_2.

$\bar{v}_i$ is the partial molar volume of component i at infinite dilution ($cm^3 \cdot mol^{-1}$) that is computed by the following correlation (Duan and Sun 2003):

$$\bar{v}_{CO_2} = -47.75418 + 4.336154 \times 10^{-1}T - 5.945771 \times 10^{-4}T^2 \quad (2.80)$$

where

T temperature, K.

References

Adibhatia B, Sun X, Mohanty K (2005) Numerical studies of oil production from initially oil-wet fracture blocks by surfactant brine imbibition. In: Paper presented at the SPE international improved oil recovery conference in Asia Pacific, Kuala Lumpur, Malaysia, 5–6 Dec 2005

Adyani WN, Wan Daud WA, Darman NB, Memon AI, Khan IA, Jamaluddin AKM (2011) A systematic approach to evaluate asphaltene precipitation during CO_2 injection. In: Paper presented at the SPE enhanced oil recovery conference, Kuala Lumpur, Malaysia, 19–21 July 2011. https://doi.org/10.2118/143903-ms

Ahmadi K, Johns RT (2011) Multiple-mixing-cell method for MMP calculations. SPE J 16(04):733–742. https://doi.org/10.2118/116823-PA

Ahmadi Y, Eshraghi SE, Bahrami P, Hasanbeygi M, Kazemzadeh Y, Vahedian A (2015) Comprehensive water-alternating-gas (WAG) injection study to evaluate the most effective method based on heavy oil recovery and asphaltene precipitation tests. J Petrol Sci Eng 133:123–129. https://doi.org/10.1016/j.petrol.2015.05.003

Ahmadloo F, Asghari K, Jamaloei BY (2009) Experimental and theoretical studies of three phase relative permeability. In: Paper presented at the SPE annual technical conference and exhibition, New Orleans, Louisiana, 4–7 Oct 2009

Ahmed T (2010) Reservoir engineering handbook. Elsevier Science, United Kingdom

Akbarzadeh K, Hammami A, Kharrat A, Zhang D, Allenson S, Creek J, Kabir S, Jamaluddin A, Marshall AG, Rodgers RP (2007) Asphaltenes—problematic but rich in potential. Oilfield Rev 19(2):22–43

Al-Aulaqi T, Grattoni C, Fisher Q, Musina Z, Al-Hinai S (2011) Effect of temperature, oil asphaltene content, and water salinity on wettability alteration. In: Paper presented at the SPE/DGS Saudi Arabia section technical symposium and exhibition, 15–18 May 2011. https://doi.org/10.2118/149071-ms

Al-Maamari RS, Buckley JS (2003) Asphaltene precipitation and alteration of wetting: the potential for wettability changes during oil production. SPE Reserv Eval Eng 6(04):210–214

Alvarado V, Manrique E (2010) Enhanced oil recovery: an update review. Energies (Basel) 3(9):1529–1575

Amao AM, Siddiqui S, Menouar H (2012) A new look at the minimum miscibility pressure (MMP) determination from slimtube measurements. In: Paper presented at the SPE improved oil recovery symposium, Tulsa, Oklahoma, USA, 14–18 April 2012. https://doi.org/10.2118/153383-ms

Aziz AK (2011) Impact of wettability alteration on recovery factor. In: Paper presented at the SPE/DGS Saudi Arabia section technical symposium and exhibition, Al-Khobar, Saudi Arabia, 15–18 May 2011

Aziz K, Settari A (1979) Petroleum reservoir simulation. Applied Science Publishers, London

Beygi MR, Delshad M, Pudugramam VS, Pope GA, Wheeler MF (2015) Novel three-phase compositional relative permeability and three-phase hysteresis models. SPE J 20(01):21–34

Buckley JS (1996) Mechanisms and consequences of wettability alteration by crude oils. Dissertation, Heriot-Watt University

Buckley JS, Liu Y, Xie X, Morrow NR (1997) Asphaltenes and crude oil wetting-the effect of oil composition. SPE J 2(02):107–119

Buckley JS, Wang J, Creek JL (2007) Solubility of the least-soluble asphaltenes. In: Mullins OC, Sheu EY, Hammami A, Marshall AG (eds) Asphaltenes, heavy oils, and petroleomics. Springer, New York, pp 401–437

Cao M, Gu Y (2013) Oil recovery mechanisms and asphaltene precipitation phenomenon in immiscible and miscible CO_2 flooding processes. Fuel 109:157–166. https://doi.org/10.1016/j.fuel.2013.01.018

Carlson FM (1981) Simulation of relative permeability hysteresis to the nonwetting phase. In: Paper presented at the SPE annual technical conference and exhibition, San Antonio, Texas, 4–7 Oct 1981

Chapman WG, Gubbins KE, Jackson G, Radosz M (1989) SAFT: equation-of-state solution model for associating fluids. Fluid Phase Equilib 52:31–38. https://doi.org/10.1016/0378-3812(89)80308-5

Chapman WG, Gubbins KE, Jackson G, Radosz M (1990) New reference equation of state for associating liquids. Ind Eng Chem Res 29(8):1709–1721. https://doi.org/10.1021/ie00104a021

Chapman WG, Jackson G, Gubbins KE (1988) Phase equilibria of associating fluids. Mol Phys 65(5):1057–1079. https://doi.org/10.1080/00268978800101601

Chen Y, Wu S, Zhou D, Chawathe A, Inouye A (2017) Impact of relative permeability hysteresis on water-alternating-gas (WAG) injectivity: modeling and experimental study. In: Paper presented at the SPE annual technical conference and exhibition, San Antonio, Texas, USA, 9–11 Oct 2017. https://doi.org/10.2118/187425-ms

Cho J, Kim TH, Chang N, Lee KS (2019) Effects of asphaltene deposition-derived formation damage on three-phase hysteretic models for prediction of coupled CO_2 enhanced oil recovery and storage performance. J Petrol Sci Eng 172:988–997. https://doi.org/10.1016/j.petrol.2018.09.006

Cho J, Park SS, Jeong MS, Lee KS (2015) Compositional modeling for optimum design of water-alternating CO_2-LPG EOR under complicated wettability conditions. J Chem 2015:9. https://doi.org/10.1155/2015/604103

Christiansen RL, Haines HK (1987) Rapid measurement of minimum miscibility pressure with the rising-bubble apparatus. SPE Reserv Eng 2(04):523–527. https://doi.org/10.2118/13114-PA

Chung TH (1992) Thermodynamic modeling for organic solid precipitation. In: Paper presented at the SPE annual technical conference and exhibition, Washington, D.C., 4–7 Oct 1992. https://doi.org/10.2118/24851-ms

Cimino R, Correra S, Sacomani PA, Carniani C (1995) Thermodynamic modelling for prediction of asphaltene deposition in live oils. In: Paper presented at the SPE international symposium on oilfield chemistry, San Antonio, Texas, 14–17 Feb 1995. https://doi.org/10.2118/28993-ms

CMG (2018) GEM user guide. Computer Modelling Group Ltd., Calgary, Alberta

Darabi H, Sepehrnoori K, Kalaei MH (2012) Modeling of wettability alteration due to asphaltene deposition in oil reservoirs. In: Paper presented at the SPE annual technical conference and exhibition, San Antonio, Texas, USA, 8–10 Oct 2012. https://doi.org/10.2118/159554-ms

Darabi H, Shirdel M, Kalaei MH, Sepehrnoori K (2014) Aspects of modeling asphaltene deposition in a compositional coupled wellbore/reservoir simulator. In: Paper presented at the SPE improved oil recovery symposium, Tulsa, Oklahoma, USA, 12 April 2014. https://doi.org/10.2118/169121-ms

Delshad M, Pope GA (1989) Comparison of the three-phase oil relative permeability models. Transp Porous Media 4(1):59–83

Dietrich JK, Bondor PL (1976) Three-phase oil relative permeability models. In: Paper presented at the SPE annual fall technical conference and exhibition, New Orleans, Louisiana, 3–6 Oct 1976. https://doi.org/10.2118/6044-ms

Duan Z, Sun R (2003) An improved model calculating CO_2 solubility in pure water and aqueous NaCl solutions from 273 to 533 K and from 0 to 2000 bar. Chem Geol 193(3–4):257–271

Edmister WC, Lee BI (1984) Applied hydrocarbon thermodynamics. Gulf Professional Publishing, Houston, Texas

Ekundayo JM, Ghedan SG (2013) Minimum miscibility pressure measurement with slim tube apparatus—how unique is the value? In: Paper presented at the SPE reservoir characterization and simulation conference and exhibition, Abu Dhabi, UAE, 16 Sept 2013. https://doi.org/10.2118/165966-ms

Escrochi M, Nabipour M, Ayatollahi SS, Mehranbod N (2008) Wettability alteration at elevated temperatures: the consequences of asphaltene precipitation. In: Paper presented at the SPE international symposium and exhibition on formation damage control, 13–15 Feb 2008

Fatemi SM, Sohrabi M (2012) Experimental and theoretical investigation of water/gas relative permeability hysteresis: applicable to water alternating gas (WAG) injection and gas storage processes. In: Paper presented at the Abu Dhabi international petroleum conference and exhibition, Abu Dhabi, UAE, 11–14 Nov 2012

Fayers FJ, Matthews J (1984) Evaluation of normalized Stone's methods for estimating three-phase relative permeabilities. SPE J 24(02):224–232

Fayers FJ (1989) Extension of Stone's method 1 and conditions for real characteristics in three-phase flow. SPE Reserv Eng 4(04):437–445. https://doi.org/10.2118/16965-PA

Fayers FJ, Foakes AP, Lin CY, Puckett DA (2000) An improved three phase flow model incorporating compositional variance. In: Paper presented at the SPE/DOE improved oil recovery symposium, Tulsa, Oklahoma, 3–5 April 2000. https://doi.org/10.2118/59313-ms

Flory PJ (1942) Thermodynamics of high polymer solutions. J Chem Phys 10(1):51–61

Ghorbani M, Momeni A, Safavi S, Gandomkar A (2014) Modified vanishing interfacial tension (VIT) test for CO_2–oil minimum miscibility pressure (MMP) measurement. J Natl Gas Sci Eng 20:92–98. https://doi.org/10.1016/j.jngse.2014.06.006

Golkari A, Riazi M (2017) Experimental investigation of miscibility conditions of dead and live asphaltenic crude oil–CO_2 systems. J Petrol Explor Prod Technol 7(2):597–609. https://doi.org/10.1007/s13202-016-0280-4

Gonzalez DL, Hirasaki GJ, Creek J, Chapman WG (2007) Modeling of asphaltene precipitation due to changes in composition using the perturbed chain statistical associating fluid theory equation of state. Energy Fuels 21(3):1231–1242

Gonzalez DL, Ting PD, Hirasaki GJ, Chapman WG (2005) Prediction of asphaltene instability under gas injection with the PC-SAFT equation of state. Energy Fuels 19(4):1230–1234

Gonzalez DL, Vargas FM, Hirasaki GJ, Chapman WG (2008) Modeling study of CO_2-induced asphaltene precipitation. Energy Fuels 22(2):757–762. https://doi.org/10.1021/ef700369u

Gonzalez K, Nasrabadi H, Barrufet M (2017) Modeling asphaltene precipitation in a compositional reservoir simulator using three-phase equilibrium. J Petrol Sci Eng 154:602–611. https://doi.org/10.1016/j.petrol.2016.09.010

Gonzalez Rodriguez DL (2008) Modeling of asphaltene precipitation and deposition tendency using the PC-SAFT equation of state. Dissertation, Rice University

Green DW, Willhite GP (1998) Enhanced oil recovery. Society of Petroleum Engineers, Richardson, Texas

Gross J, Sadowski G (2001) Perturbed-chain SAFT: an equation of state based on a perturbation theory for chain molecules. Ind Eng Chem Res 40(4):1244–1260. https://doi.org/10.1021/ie0003887

Gu Y, Hou P, Luo W (2013) Effects of four important factors on the measured minimum miscibility pressure and first-contact miscibility pressure. J Chem Eng Data 58(5):1361–1370. https://doi.org/10.1021/je4001137

Gupta AK (1986) A model for asphaltene flocculation using an equation of state. Dissertation, University of Calgary

Hajilary N, Shahmohammadi A (2018) New permeability model for gel coated porous media with radial flow. J Appl Fluid Mech 11:397–404. https://doi.org/10.29252/jafm.11.02.27874

Harvey AH (1996) Semiempirical correlation for Henry's constants over large temperature ranges. AIChE J 42(5):1491–1494

Hildebrand JH, Scott RL (1950) Solutions of nonelectrolytes. Annu Rev Phys Chem 1(1):75–92. https://doi.org/10.1146/annurev.pc.01.100150.000451

Hirschberg A, deJong LNJ, Schipper BA, Meijer JG (1984) Influence of temperature and pressure on asphaltene flocculation. SPE J 24(03):283–293. https://doi.org/10.2118/11202-PA

Holm LW, Josendal VA (1974) Mechanisms of oil displacement by carbon dioxide. J Petrol Technol 26(12):1427–1438. https://doi.org/10.2118/4736-PA

Holtz M (2002) Residual gas saturation to aquifer influx: a calculation method for 3-D computer reservoir model construction. In: Paper presented at the SPE gas technology symposium, Calgary, Alberta, 30 April 2002. https://doi.org/10.2118/75502-ms

Huggins ML (1941) Solutions of long chain compounds. J Chem Phys 9(5):440–440

Hustad OS, Holt T (1992) Gravity stable displacement of oil by hydrocarbon gas after waterflooding. In: Paper presented at the SPE/DOE enhanced oil recovery symposium, Tulsa, Oklahoma, 22–24 April 1992. https://doi.org/10.2118/24116-ms

Iglauer S, Fernø M, Shearing P, Blunt M (2012) Comparison of residual oil cluster size distribution, morphology and saturation in oil-wet and water-wet sandstone. J Colloid Interface Sci 375(1):187–192

Jamaluddin AKM, Creek J, Kabir CS, McFadden JD, Apos, Cruz D, Joseph MT, Joshi N, Ross B (2001) A comparison of various laboratory techniques to measure thermodynamic asphaltene instability. In: Paper presented at the SPE Asia Pacific improved oil recovery conference, Kuala Lumpur, Malaysia, 6–9 Oct 2001. https://doi.org/10.2118/72154-ms

Jaubert J-N, Wolff L, Neau E, Avaullee L (1998) A very simple multiple mixing cell calculation to compute the minimum miscibility pressure whatever the displacement mechanism. Ind Eng Chem Res 37(12):4854–4859. https://doi.org/10.1021/ie980348r

Jha KNN (1986) A laboratory study of heavy oil recovery with carbon dioxide. J Can Pet Technol 25(02):11. https://doi.org/10.2118/86-02-03

Jhaveri BS, Youngren GK (1988) Three-parameter modification of the Peng–Robinson equation of state to improve volumetric predictions. SPE Reserv Eng 3(03):1033–1040. https://doi.org/10.2118/13118-PA

Johns RT, Dindoruk B, Orr FM (1993) Analytical theory of combined condensing/vaporizing gas drives. SPE Adv Technol Ser 1(02):7–16. https://doi.org/10.2118/24112-PA

Kohse BF, Nghiem LX (2004) Modelling asphaltene precipitation and deposition in a compositional reservoir simulator. In: Paper presented at the SPE/DOE symposium on improved oil recovery, Tulsa, Oklahoma, 17–21 April 2004. https://doi.org/10.2118/89437-ms

Kohse BF, Nghiem LX, Maeda H, Ohno K (2000) Modelling phase behaviour including the effect of pressure and temperature on asphaltene precipitation. In: Paper presented at the SPE Asia Pacific oil and gas conference and exhibition, Brisbane, Australia, 16–18 Oct 2000

Kokal S, Maini BB (1990) An improved model for estimating three-phase oil-water-gas relative permeabilities from two-phase oil-water and oil-gas data. J Can Petrol Technol 29(02)

Kontogeorgis GM, Voutsas EC, Yakoumis IV, Tassios DP (1996) An equation of state for associating fluids. Ind Eng Chem Res 35(11):4310–4318. https://doi.org/10.1021/ie9600203

Kraiwattanawong K, Fogler HS, Gharfeh SG, Singh P, Thomason WH, Chavadej S (2007) Thermodynamic solubility models to predict asphaltene instability in live crude oils. Energy Fuels 21(3):1248–1255. https://doi.org/10.1021/ef060386k

Lake LW, Johns R, Rossen B (2014) Fundamentals of enhanced oil recovery. Society of Petroleum Engineers, Richardson, Texas

Land CS (1968) Calculation of imbibition relative permeability for two-and three-phase flow from rock properties. SPE J 8(02):149–156

Larsen J, Skauge A (1998) Methodology for numerical simulation with cycle-dependent relative permeabilities. SPE J 3(02):163–173

Le Van S, Chon BH (2017) Effects of salinity and slug size in miscible CO_2 water-alternating-gas core flooding experiments. J Ind Eng Chem 52:99–107. https://doi.org/10.1016/j.jiec.2017.03.030

Li YK, Nghiem LX (1986) Phase equilibria of oil, gas and water/brine mixtures from a cubic equation of state and Henry's law. Can J Chem Eng 64(3):486–496

Li Z, Firoozabadi A (2010) Cubic-plus-association equation of state for asphaltene precipitation in live oils. Energy Fuels 24(5):2956–2963. https://doi.org/10.1021/ef9014263

Mahmoudvand S, Shahsavani B, Parsaei R, Malayeri MR (2019) Prediction of asphaltene precipitation upon injection of various gases at near-wellbore conditions: a simulation study using PC-SAFT EOS. Oil Gas Sci Technol—Revue d IFP Energies nouvelles 74(63)

Mahzari P, Sohrabi M (2016) An improved approach for estimation of flow and hysteresis parameters applicable to WAG experiments. In: Paper presented at the Abu Dhabi international petroleum exhibition and conference, Abu Dhabi, UAE, 7–10 Nov 2016. https://doi.org/10.2118/183321-ms

Mahzari P, Sohrabi M (2017) An improved approach for estimation of flow and hysteresis parameters applicable to WAG experiments. Fuel 197:359–372

Metcalfe RS (1982) Effects of impurities on minimum miscibility pressures and minimum enrichment levels for CO_2 and rich-gas displacements. SPE J 22(02):219–225. https://doi.org/10.2118/9230-PA

Mohebbinia S, Sepehrnoori K, Johns RT, Kazemi Nia Korrani A (2014) Simulation of asphaltene precipitation during gas injection using PC-SAFT EOS. In: Paper presented at the SPE annual technical conference and exhibition, Amsterdam, The Netherlands, 27–29 Oct 2014. https://doi.org/10.2118/170697-ms

Monroe WW, Silva MK, Larson LL, Orr FM (1990) Composition paths in four-component systems: effect of dissolved methane on 1D CO_2 flood performance. SPE Reserv Eng 5(03):423–432. https://doi.org/10.2118/16712-PA

Morrow NR (1990) Wettability and its effect on oil recovery. J Petrol Technol 42(12):1476–1484

Mullins OC (2005) Molecular structure and primary aggregation of asphaltenes and petroleomics. In: Paper presented at the SPE annual technical conference and exhibition, Dallas, Texas, 9–12 Oct 2005. https://doi.org/10.2118/95801-ms

Neau E, Avaullée L, Jaubert JN (1996) A new algorithm for enhanced oil recovery calculations. Fluid Phase Equilib 117(1):265–272. https://doi.org/10.1016/0378-3812(95)02962-1

Nghiem L, Hassam M, Nutakki R, George A (1993) Efficient modelling of asphaltene precipitation. In: Paper presented at the SPE annual technical conference and exhibition, Houston, Texas, 3–6 Oct 1993. https://doi.org/10.2118/26642-ms

Nghiem LX, Coombe DA (1997) Modelling asphaltene precipitation during primary depletion. SPE J 2(02):170–176

Nghiem LX, Kohse BF, Ali S, Doan Q (2000) Asphaltene precipitation: phase behaviour modelling and compositional simulation. In: Paper presented at the SPE Asia Pacific conference on integrated modelling for asset management, Yokohama, Japan, 25–26 April 2000. https://doi.org/10.2118/59432-ms

Nor-Azlan N, Adewumi MA (1993) Development of asphaltene phase equilibria predictive model. In: Paper presented at the SPE eastern regional meeting, Pittsburgh, Pennsylvania, 2–4 Nov 1993. https://doi.org/10.2118/26905-ms

Orr FM (2007) Theory of gas injection processes, vol 5. Tie-Line Publications, Copenhagen, Denmark

Oskui G-RP, Jumaa MA, Abuhained W (2009) Laboratory investigation of asphaltene precipitation problems during CO_2/hydrocarbon injection project for EOR application in Kuwaiti reservoirs. In: Paper presented at the Kuwait international petroleum conference and exhibition, Kuwait City, Kuwait, 14–16 Dec 2009. https://doi.org/10.2118/126267-ms

Pan H, Firoozabadi A (1997) Thermodynamic micellization model for asphaltene precipitation from reservoir crudes at high pressures and temperatures. In: Paper presented at the SPE annual technical conference and exhibition, San Antonio, Texas, 5–8 Oct 1997. https://doi.org/10.2118/38857-ms

Panuganti SR, Tavakkoli M, Vargas FM, Gonzalez DL, Chapman WG (2013) SAFT model for upstream asphaltene applications. Fluid Phase Equilib 359:2–16

Pejic D, Maini BB (2003) Three-phase relative permeability of petroleum reservoirs. In: Paper presented at the SPE Latin American and Caribbean petroleum engineering conference, Port-of-Spain, Trinidad and Tobago, 27–30 April 2003

Péneloux A, Rauzy E, Fréze R (1982) A consistent correction for Redlich-Kwong-Soave volumes. Fluid Phase Equilib 8(1):7–23. https://doi.org/10.1016/0378-3812(82)80002-2

Peng D-Y, Robinson DB (1976) A new two-constant equation of state. Ind Eng Chem Fundam 15(1):59–64. https://doi.org/10.1021/i160057a011

Pentland CH, Itsekiri E, Al-Mansoori S, Iglauer S, Bijeljic B, Blunt MJ (2010) Measurement of nonwetting-phase trapping in sandpacks. SPE J 15(02):274–281

Pitzer KS, Lippmann DZ, Curl RF, Huggins CM, Petersen DE (1955) The volumetric and thermodynamic properties of fluids. II. Compressibility factor, vapor pressure and entropy of vaporization. J Am Chem Soc 77(13):3433–3440. https://doi.org/10.1021/ja01618a002

Randall TE, Bennion DB (1988) Recent developments in slim tube testing for hydrocarbon-miscible flood (HCMF) solvent design. J Can Pet Technol 27(06):13. https://doi.org/10.2118/88-06-02

Randall TE, Bennion DB (1989) Laboratory factors influencing slim tube test results. J Can Pet Technol 28(04):12. https://doi.org/10.2118/89-04-03

Rao DN (1997) A new technique of vanishing interfacial tension for miscibility determination. Fluid Phase Equilib 139(1):311–324. https://doi.org/10.1016/S0378-3812(97)00180-5

Redlich O, Kwong JNS (1949) On the thermodynamics of solutions. V. An equation of state. Fugacities of gaseous solutions. Chem Rev 44(1):233–244. https://doi.org/10.1021/cr60137a013

Reid RC, Reid RD, Prausnitz JM, Sherwood TK, Poling BE (1977) The properties of gases and liquids. McGraw-Hill, New York

Reis JC, Acock AM (1994) Permeability reduction models for the precipitation of inorganic solids in Berea sandstone. In situ 18(3):347–368

Robinson DB, Peng DY (1978) The characterization of the heptanes and heavier fractions for the GPA Peng–Robinson programs. Gas Processors Association, Tulsa, Oklahoma

Rojas GA, Ali SMF (1988) Dynamics of subcritical CO_2/brine floods for heavy-oil recovery. SPE Reserv Eng 3(01):35–44. https://doi.org/10.2118/13598-PA

Rommerskirchen R, Nijssen P, Bilgili H, Sottmann T (2016) Reducing the miscibility pressure in gas injection oil recovery processes. In: Paper presented at the Abu Dhabi international petroleum exhibition and conference, Abu Dhabi, UAE, 7 Nov 2016. https://doi.org/10.2118/183389-ms

Sabeti M, Rahimbakhsh A, Nikookar M, Mohammadi AH (2015) Estimation of asphaltene precipitation and equilibrium properties of hydrocarbon fluid phases using the PC-SAFT equation of state. J Mol Liq 209:447–460

Salathiel R (1973) Oil recovery by surface film drainage in mixed-wettability rocks. J Petrol Technol 25(10):1216–1224

Saul A, Wagner W (1987) International equations for the saturation properties of ordinary water substance. J Phys Chem Ref Data 16(4):893–901. https://doi.org/10.1063/1.555787

Schou Pedersen K, Hasdbjerg C (2007) PC-SAFT equation of state applied to petroleum reservoir fluids. In: Paper presented at the SPE annual technical conference and exhibition, California, U.S.A., 11–14 Nov 2007. https://doi.org/10.2118/110483-ms

Shahrokhi O, Fatemi M, Sohrabi M, Ireland S, Ahmed K (2014) Assessment of three phase relative permeability and hysteresis models for simulation of water-alternating-gas (WAG) injection in water-wet and mixed-wet systems. In: Paper presented at the SPE improved oil recovery symposium, Tulsa, Oklahoma, USA, 12–16 April 2014. https://doi.org/10.2118/169170-ms

Shirani B, Nikazar M, Mousavi-Dehghani SA (2012) Prediction of asphaltene phase behavior in live oil with CPA equation of state. Fuel 97:89–96. https://doi.org/10.1016/j.fuel.2012.02.016

Soave G (1972) Equilibrium constants from a modified Redlich–Kwong equation of state. Chem Eng Sci 27(6):1197–1203. https://doi.org/10.1016/0009-2509(72)80096-4

Spiteri EJ, Juanes R (2006) Impact of relative permeability hysteresis on the numerical simulation of WAG injection. J Petrol Sci Eng 50(2):115–139

Spiteri EJ, Juanes R, Blunt MJ, Orr FM (2008) A new model of trapping and relative permeability hysteresis for all wettability characteristics. SPE J 13(03):277–288

Srivastava RK, Huang SS, Dong M (1999) Asphaltene deposition during CO_2 flooding. SPE Prod Facil 14(04):235–245. https://doi.org/10.2118/59092-PA

Stalkup FI (1983) Miscible displacement. Society of Petroleum Engineers of AIME, Dallas, New York

Stone H (1970) Probability model for estimating three-phase relative permeability. J Petrol Technol 22(02):214–218

Stone H (1973) Estimation of three-phase relative permeability and residual oil data. J Can Petrol Technol 12(04)

Subramanian S, Simon S, Sjöblom J (2016) Asphaltene precipitation models: a review. J Dispers Sci Technol 37(7):1027–1049. https://doi.org/10.1080/01932691.2015.1065418

Suzanne K, Hamon G, Billiotte J, Trocme V (2003) Experimental relationships between residual gas saturation and initial gas saturation in heterogeneous sandstone reservoirs. In: Paper presented at the SPE annual technical conference and exhibition, Denver, Colorado, 5–8 Oct 2003

Thomas LK, Coats KH. (1992). Stones kro methods and modifications. Society of Petroleum Engineers, p. 3

Treiber L, Owens W (1972) A laboratory evaluation of the wettability of fifty oil-producing reservoirs. SPE J 12(06):531–540

Uetani T (2014) Wettability alteration by asphaltene deposition: a field example. In: Paper presented at the Abu Dhabi international petroleum exhibition and conference, Abu Dhabi, UAE, 10–13 Nov 2014

Van der Waals JD (1873) On the continuity of the gas and liquid state. Dissertation, University of Leiden

Vargas FM, Gonzalez DL, Hirasaki GJ, Chapman WG (2009) Modeling asphaltene phase behavior in crude oil systems using the perturbed chain form of the statistical associating fluid theory (PC-SAFT) equation of state. Energy Fuels 23(3):1140–1146

Victorov AI, Firoozabadi A (1996) Thermodynamic micellizatin model of asphaltene precipitation from petroleum fluids. AIChE J 42(6):1753–1764

Wang S, Civan F (2001) Productivity decline of vertical and horizontal wells by asphaltene deposition in petroleum reservoirs. In: Paper presented at the SPE international symposium on oilfield chemistry, Houston, Texas, 13–16 Feb 2001

Wang X, Gu Y (2011) Oil recovery and permeability reduction of a tight sandstone reservoir in immiscible and miscible CO_2 flooding processes. Ind Eng Chem Res 50(4):2388–2399. https://doi.org/10.1021/ie1016046

Wang Y, Orr FM (1997) Analytical calculation of minimum miscibility pressure. Fluid Phase Equilib 139(1):101–124. https://doi.org/10.1016/S0378-3812(97)00179-9

Wertheim MS (1984a) Fluids with highly directional attractive forces. I. Statistical thermodynamics. J Stat Phys 35(1):19–34. https://doi.org/10.1007/bf01017362

Wertheim MS (1984b) Fluids with highly directional attractive forces. II. Thermodynamic perturbation theory and integral equations. J Stat Phys 35(1):35–47. https://doi.org/10.1007/bf01017363

Wertheim MS (1986a) Fluids with highly directional attractive forces. III. Multiple attraction sites. J Stat Phys 42(3):459–476. https://doi.org/10.1007/bf01127721

Wertheim MS (1986b) Fluids with highly directional attractive forces. IV. Equilibrium polymerization. J Stat Phys 42(3):477–492. https://doi.org/10.1007/bf01127722

Whitson CH, Brulé MR (2000) Phase behavior. Society of Petroleum Engineers, Richardson, Texas

Wilson G (1968) A modified Redlich-Kwong EOS, application to general physical data calculations. In: Paper presented at the annual AIChE national meeting, Cleveland, 4–7 May 1968

Wu RS, Batycky JP (1990) Evaluation of miscibility from slim tube tests. J Can Pet Technol 29(06):9. https://doi.org/10.2118/90-06-06

Yellig WF, Metcalfe RS (1980) Determination and prediction of CO_2 minimum miscibility pressures (includes associated paper 8876). J Petrol Technol 32(01):160–168. https://doi.org/10.2118/7477-PA

Yuan C, Pope G (2012) A new method to model relative permeability in compositional simulators to avoid discontinuous changes caused by phase-identification problems. SPE J 17(04):1221–1230. https://doi.org/10.2118/142093-PA

Zhao G, Adidharma H, Towler BF, Radosz M (2006) Minimum miscibility pressure prediction using statistical associating fluid theory: two- and three-phase systems. In: Paper presented at the SPE annual technical conference and exhibition, San Antonio, Texas, USA, 24–27 Sept 2006. https://doi.org/10.2118/102501-ms

Zúñiga-Hinojosa MA, Justo-García DN, Aquino-Olivos MA, Román-Ramírez LA, García-Sánchez F (2014) Modeling of asphaltene precipitation from n-alkane diluted heavy oils and bitumens using the PC-SAFT equation of state. Fluid Phase Equilib 376:210–224

Chapter 3
Numerical Modeling of CO_2 EOR

Abstract The methodologies for compositional modeling of CO_2 EOR process is described by using the theories from Chap. 2. Fluid modeling for MMP estimation and asphaltene precipitation is performed before conducting reservoir simulation. Weyburn fluid and Burke Oil 1 are used for fluid modeling. In the fluid modeling, the addition of LPG contributes to reduce MMP but to accelerate asphaltene precipitation. Then, reservoir simulation for CO_2 and CO_2-LPG EOR is introduced to explain the effect of CO_2 and intermediate hydrocarbons on miscibility and its performance by using the developed fluid model of Weyburn oil. LPG addition into CO_2 stream improves not only displacement efficiency but also vertical sweep efficiency which results in enhancing oil production. The base models of CO_2 and CO_2-LPG WAG processes are developed by using Burke Oil 1. WAG is an effective injection strategy to increase the performance of EOR and residual and solubility trapping for CCS. Asphaltene deposition may occur during WAG process everywhere in the reservoir, and LPG addition accelerates the amount of asphaltene deposition. Although LPG addition improves the WAG performance by enhancing displacement efficiency with the effect of miscibility improvement, the effects can be overestimated if formation damages by asphaltene deposition are not considered.

Keywords Fluid modeling · Weyburn fluid · Burke oil 1 · Miscibility · Asphaltene deposition · Liquefied petroleum gas (LPG) · Water alternating gas (WAG)

3.1 Fluid Modeling

3.1.1 MMP Estimation

MMP is the most significant factor to confirm the efficiency of CO_2 EOR by establishing the miscibility condition which results in a higher ultimate oil recovery. A number of researches have been conducted to estimate MMP by experimental or non-experimental ways for decades (Yellig and Metcalfe 1980; Stalkup 1983; Whitson and Michelsen 1989; Elsharkawy et al. 1992; Dorshow 1995; Rao 1997;

K. S. Lee et al., *CO_2 Storage Coupled with Enhanced Oil Recovery*,
https://doi.org/10.1007/978-3-030-41901-1_3

Orr 2007; Abedini et al. 2014; Sharbatian et al. 2018; Zhang et al. 2018). The effect of impurities in injected CO_2 stream on MMP has been also studied. Conducting slim tube tests of Weyburn oil to estimate MMP with pure CO_2 and mixtures, Huang and Dyer (1992) confirmed that the addition of 9.9 mol% CH_4 in CO_2 stream increases the MMP by 4.7 MPa. Moreover, the addition of 5.1 mol% of N_2 and 5.1 mol% of CH_4 rises MMP by 8.4 MPa, which indicates that CH_4 and N_2 increases MMP and CH_4 is more sensitive. Zhang et al. (2004) also performed the rising bubble experiment to determine the MMP of pure CO_2 and impurities for Weyburn fluid. They concluded that the addition of an intermediate hydrocarbon gas such as propane reduces MMP, so propane is a more effective agent than ethane. Sayegh et al. (2007) studied the impacts of H_2S and pressure depletion on the MMP of CO_2, resulting in an almost linear reduction of MMP with increasing amounts of H_2S in the injected CO_2 stream. Talbi and Maini (2008) conducted experimental research on displacement efficiency improvement such as oil swelling, viscosity reduction, interfacial tension reduction, and oil recovery improvement that resulted from injecting solvents into CO_2 stream. Teklu et al. (2014) focused on the relationship among pore confinement, permeability, and MMP in shale reservoirs. They confirmed MMP reduction effect by liquefied petroleum gas (LPG) addition into CO_2 stream in various reservoir scenarios by using simulation model.

For the numerical modeling of CO_2 EOR process, fluid modeling considering MMP should be preceded by reservoir simulation in dynamic condition. The data of Weyburn W3, which is widely used for CO_2 EOR simulation studies, is utilized in this fluid modeling section (Srivastava et al. 2000). Weyburn reservoir, located in southeast Saskatchewan and operated by PanCanadian Petroleum Ltd., has reached its economic limit of production by water flooding, so the reservoir is a target for CO_2 miscible flooding for EOR (Srivastava et al. 1999). Therefore, W3 is modeled as a preliminary step to investigate the reliability of fluid modeling methodologies by matching oil properties and MMP to experimental data before developing integrated model of CO_2 EOR. The saturation pressure of W3 is 713 psi at reservoir temperature of 145 °F, and the oil density and viscosity at saturation pressure are 50.3 lb/ft^3 and 1.76 cp, and MMP with CO_2 is 2,059 psi at reservoir temperature (Srivastava et al. 2000).

Reservoir compositional simulation usually requires large CPU time to describe the phase and volumetric behavior at each time step. Therefore, many researches have been conducted to group heavy components by the splitting and lumping methods (Lee et al. 1981; Whitson 1983; Coats 1985; Li et al. 1985; Joergensen and Stenby 1995; Alavian et al. 2014). After grouping the components, the properties of lumped components are not persisted from the pure components, so they should be used as matching parameters to experimental data. Model properties of W3 oil, such as saturation pressure, etc., are matched to experimental data by using regression method, and the matching results are shown in Table 3.1. To detect the effect of additives to CO_2 stream on EOR performance, MMPs with various LPG concentration are calculated by using multiple mixing cell method as indicated in Table 3.2. The MMP calculation result with pure CO_2 is well matched to experimental data of 2,059 psi to confirm the reliability of the fluid model. MMPs show a decreasing

Table 3.1 Comparison of properties between Weyburn W3 experimental data and fluid model

Parameters	W3	Fluid model
Saturation pressure (psi)	713	693
Oil density at ρ_{sat} (lb/ft^3)	50.3	50.3
Viscosity at ρ_{sat} (cp)	1.76	1.75
Formation volume factor	1.12	1.11
API°	31	34.8

Table 3.2 Calculated MMPs between Weyburn fluid and injected gas

Components	MMP (psi)
CO_2 100%	2,012
CO_2 95% + LPG 5%	1,995
CO_2 90% + LPG 10%	1,825
CO_2 85% + LPG 15%	1,820
CO_2 80% + LPG 20%	1,412
CO_2 75% + LPG 25%	1,354
CO_2 70% + LPG 30%	1,046

tendency as LPG (propane 37% and butane 63%) mole fraction increases. The oil composition and properties after matching the experimental data for the calculation by using PR EOS are indicated in Table 3.3.

Following the procedure, MMP of Burke Oil 1 (Burke et al. 1990) is estimated by multiple mixing cell method as 4,870 psi with pure CO_2. Moreover, MacLeod-Sugden correlation proposed by Reid et al. (1977) is applied to compute IFT between displacing fluid and reservoir oil to crosscheck the MMP which is estimated from

Table 3.3 Oil composition and properties of Weyburn fluid for input data into PR EOS

Component	Mole fraction	ρ_c (psi)	T_c (°F)	Acentric factor	Molecular weight
N_2	0.0207	492	−232.69	0.04	28.01
CO_2	0.0074	1,070	87.89	0.225	44.01
H_2S	0.0012	1,297	212.09	0.1	34.08
Methane	0.0749	667	−116.59	0.008	16.04
Ethane	0.0422	709	90.05	0.098	30.07
Propane	0.0785	616	205.97	0.152	44.10
Butane	0.0655	540	290.1	0.1845	58.12
Pentane	0.0459	490	377.3	0.239	72.15
C_{6-9}	0.2155	438	541.79	0.331	102.5
C_{10-17}	0.2202	293	786.46	0.584	184
C_{18-27}	0.1027	193	995.5	0.893	306.2
C_{28+}	0.1252	168	1,188.3	1.1	565.61

Table 3.4 Results of MMP calculation of Burke Oil 1 by multiple mixing cell and IFT methods with various LPG concentration (psi)

Components	Multiple mixing cell method	IFT method
CO_2	4,870	4,786
CO_2 90% + LPG 10%	4,450	4,206
CO_2 80% + LPG 20%	3,364	3,480

the previous method. IFT is measured at a sequence of pressures and the MMP is estimated to be the pressure that IFT plotted as a function of pressure extrapolates to zero (Jessen and Orr 2008). MMP is considered as pressure that IFT becomes less than 0.001 dyne/cm. In slim-tube system model, IFT is calculated, and the criterion is also decided to 0.001 dyne/cm in this example. Results of MMP computation of Burke Oil 1 by multiple mixing cell and IFT methods are indicated in Table 3.4. As the LPG concentration increases, the MMP becomes lower. The tendency is already verified by experiments and the previous Weyburn oil case (Metcalfe 1982; Teletzke et al. 2005). The detailed fluid modeling of Burke Oil 1, such as components properties considering asphaltene precipitation, is described in the following by the next section.

3.1.2 Asphaltene Modeling

The Burke Oil 1, which is widely used for simulation studies, has been utilized in this book for asphaltene modeling with pressure and composition changes (Burke et al. 1990; Fazelipour 2011; Darabi et al. 2012, 2014; Gonzalez et al. 2014; Cho et al. 2016, 2019). Burke Oil 1 is a live oil with molecular weight of 171.4, API gravity of 19, and a saturation pressure of 2,950 psi at a reservoir temperature of 212 °F (Burke et al. 1990). The experimental and computed oil properties are shown in Table 3.5. Regression method shows acceptable results that were applied for matching fluid model properties to experimental data. Moreover, the oil composition and properties of each component after matching process are indicated in Table 3.6. As the heaviest component is divided into C_{31A+} and C_{31B+} that each indicates non-precipitating and precipitating component, all properties are the same except composition and binary interaction coefficient. The amount of asphaltene precipitation is also matched to reference data of Darabi et al. (2012) as shown in Fig. 3.1. The oil volume expands

Table 3.5 The properties between Burke Oil 1 experimental data and fluid model

Parameters	Oil 1	Fluid model
Saturation pressure (psi)	2,950	2,950
Live oil molecular weight (g/mol)	171.4	174
API (°)	19	19

Table 3.6 Oil composition of Burke Oil 1 and properties of each component

Component	Mole fraction	ρ_c (psi)	T_c (°F)	Molecular weight
CO_2	0.0246	1,070	87.89	44.0
N_2	0.0057	492	−232.51	28.0
CH_4	0.3637	667	−116.59	16.0
C_2H_6	0.0347	708	90.05	30.1
C_3H_8	0.0405	616	205.97	44.1
C_4H_{10}	0.0193	544	296.15	58.1
i-C_5	0.0074	491	369.05	72.2
n-C_5	0.0083	489	385.61	72.2
FC_6	0.0162	477	453.83	86.0
C_{7-15}	0.1966	381	715.01	153.4
C_{16-25}	0.1255	235	998.15	293.3
C_{26-30}	0.0400	177	1159.79	389.5
C_{31A+}	0.0742	100	1476.59	665.6
C_{31B+}	0.0433	100	1476.59	665.6

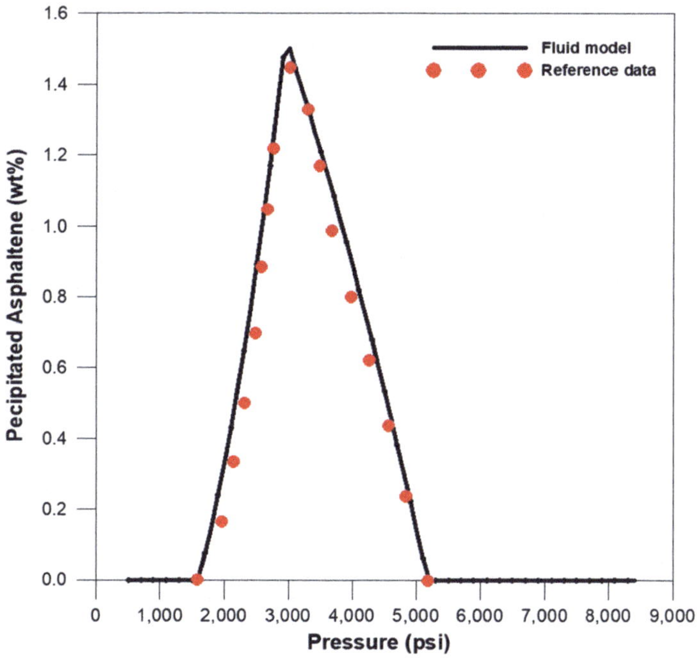

Fig. 3.1 The amount of asphaltene precipitation of Burke Oil 1 with pressure change

as pressure decreases, resulting in reduction of asphaltene solubility. It starts to precipitate when the system pressure reaches onset point pressure of 5,173 psi. The amount of asphaltene precipitation increases consistently and has the maximum value at saturation pressure of 2,950 psi. After the solution gases are released from the reservoir oil under saturation pressure, oil becomes denser and causes asphaltene re-dissolution. Therefore, asphaltene precipitation does not occur any more below offset point pressure of 1,600 psi. Asphaltene precipitation due to oil composition change by CO_2 contact changes the amount, onset, and offset pressures of asphaltene precipitation simultaneously (Darabi et al. 2014). Figure 3.2 indicates the amount of asphaltene precipitation with mole fraction change of CO_2 and oil at the reservoir pressure of 2,800 psi and temperature of 212 °F that will be conducted in the next reservoir simulation section. Asphaltene precipitation is accelerated by the increase of CO_2 mol fraction, and it starts to decrease above 0.7 of CO_2 fraction because of two-phase region (Srivastava et al. 1999; Cho et al. 2019). Moreover, LPG accelerates oil swelling resulting in decreasing the stability of asphaltene in oil phase, which causes more asphaltene precipitation compared with pure CO_2 injection case (Cho et al. 2016; Fig. 3.3).

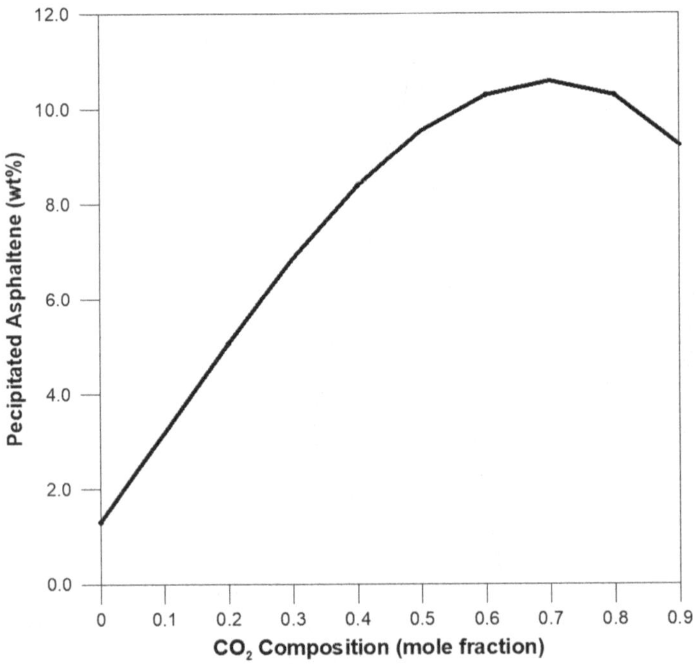

Fig. 3.2 The amount of asphaltene precipitation of Burke Oil 1 with injected CO_2 mol fraction (Cho et al. 2019)

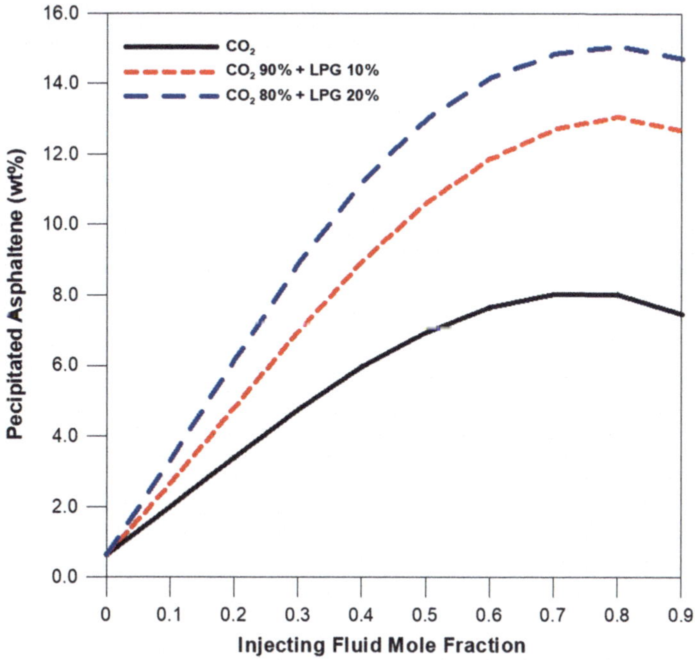

Fig. 3.3 The amount of asphaltene precipitation with various injecting fluids of CO_2 and LPG mole fraction at reservoir pressure and temperature

3.2 Reservoir Simulation in Dynamic Condition

3.2.1 CO_2-LPG EOR Model

Though miscible gas injection process has effectiveness, the application of the process is limited because of the condition that the reservoir pressure should be higher than MMP. The limitation can be overcome by the addition of LPG into the CO_2 stream, which can improve the displacement efficiency by lowering the MMP and vertical sweep efficiency by reducing viscous gravity number, resulting in oil recovery improvement (Kumar and Von Gonten 1973; Alston et al. 1985; Chugh et al. 2000; Talbi and Maini 2008; Lake 2010; Cho and Lee 2016).

The W3 fluid data of Weyburn reservoir and 2-D vertical model were used for the compositional simulation to explain the effects of LPG on displacement and sweep efficiencies simultaneously. The reservoir model is assumed to be homogeneous and discretized into $52 \times 1 \times 20$ grid blocks, and each grid block has the dimensions of 10 ft $\times$ 10 ft $\times$ 5 ft. The reservoir is in oil wet condition, and the initial and operating conditions are indicated in Table 3.7. After the water flooding is conducted for three years, CO_2 and CO_2-LPG mixture are injected for ten years. Because the MMP of pure CO_2 is lower than reservoir system pressure, the reservoir achieves

Table 3.7 The initial and operating conditions for CO_2-LPG EOR

Properties	Values
Initial pressure (psi)	2,000
Initial oil saturation	0.88
Porosity (%)	30
Permeability (md)	50
Producing pressure (psi)	1,600
Total injection (PV)	2

miscible condition when 20% of LPG is added. As shown in Fig. 3.4, the oil viscosity in the swept zone is reduced by CO_2 injection from 2.0 to 0.6 cp when 0.4 PV of CO_2 is injected, and the minimum value of oil viscosity is 0.5 cp with LPG 10% addition. The effects become much clearer as the injected LPG mole fraction increased. The addition of 20% LPG accelerates the oil viscosity reduction up to 0.3 cp, which is 50% lower than that of the CO_2 only case. In the white zone, the oil viscosity is greater than 4 cp which is higher than the original values near the wellbore because intermediate components are displaced by displacing fluid and only the heavy components comprise the residual oil. Because more oil is displaced due to added LPG, greater oil is accumulated at the front of displacing fluid as shown in Fig. 3.5. More stable and steeper displacing fluid front also indicates improved vertical sweep efficiency when 20% LPG case is compared to CO_2 100% case. The more detailed analyses and results are provided in the reference from Cho and Lee (2016).

3.3 CO_2 and CO_2-LPG WAG Model with Formation Damages

To explain the process of improved simulation models to CCS-EOR with asphaltene deposition in Chaps. 3 and 4, the Burke Oil 1 is used because it is widely applied to simulation studies with asphaltene problem. Firstly, the model for gas injection scheme of water alternating gas (WAG) is developed as a base case. When CO_2 is injected into reservoirs, it improves displacement efficiency by decreasing oil viscosity, IFT, and oil density compared to water flooding. However, gravity override due to the high density and viscosity differences between oil and CO_2 causes poor sweep efficiency. WAG is an effective injection strategy to strengthen the strength and make up for the weakness of water and CO_2 (Li et al. 2003; Righi et al. 2004; Brown et al. 2013; Jun et al. 2013). Since WAG also increases the performance of residual and solubility trapping for CCS, it is used in this example (Sifuentes et al. 2009).

The reservoir is modeled as 2-D homogeneous areal system with 33 × 33 × 1 grid blocks having size of 10 ft × 10 ft × 20 ft. Total processes last for 13 years

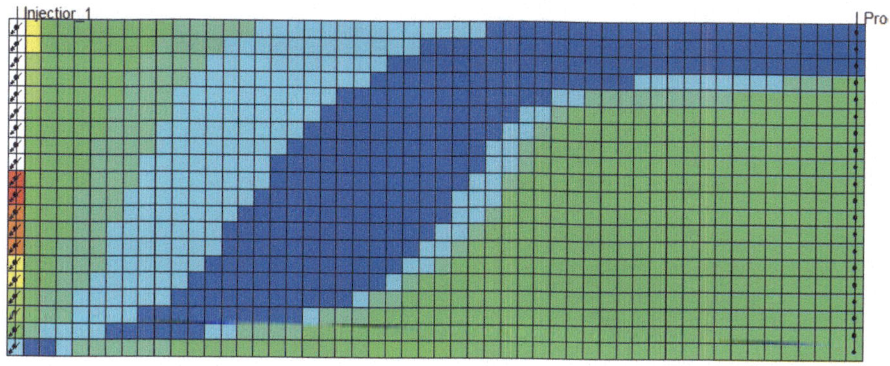

(a) CO_2 100%

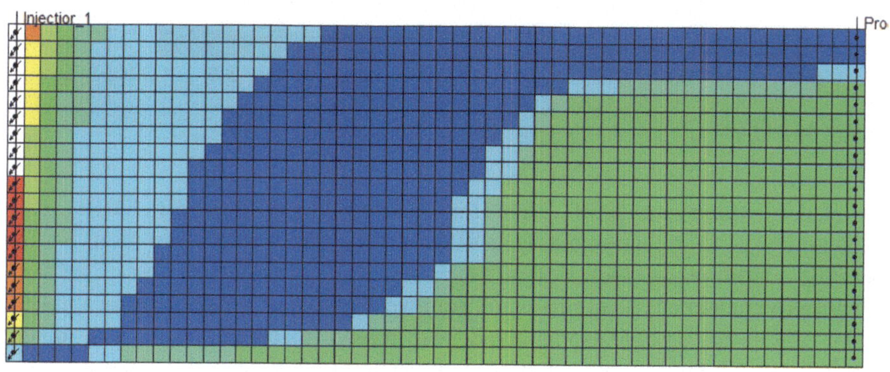

(b) CO_2 90% and LPG 10%

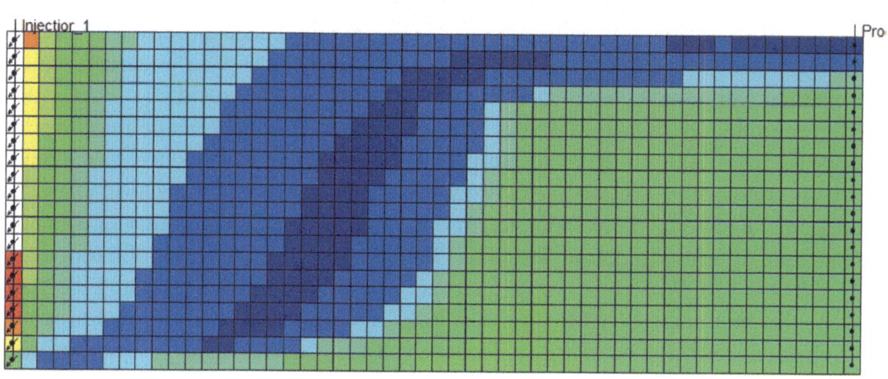

(c) CO_2 85% and LPG 15%

Fig. 3.4 Oil viscosities with **a** CO_2 100%, **b** CO_2 90% + LPG 10%, **c** CO_2 85% + LPG 15%, and **d** CO_2 80% + LPG 20% at 0.4 PV injection

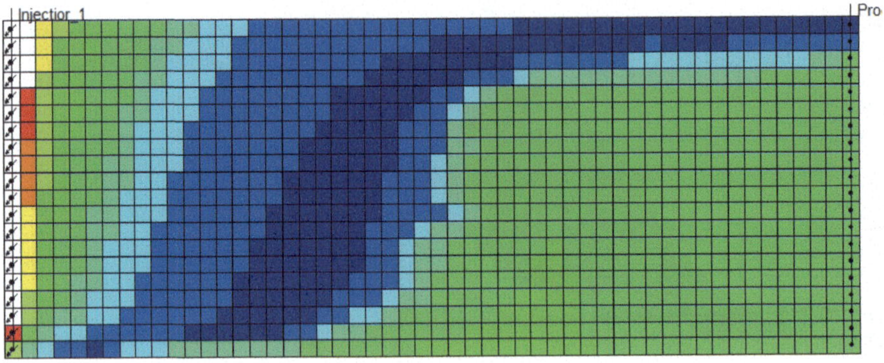

(d) CO_2 80% and LPG 20%

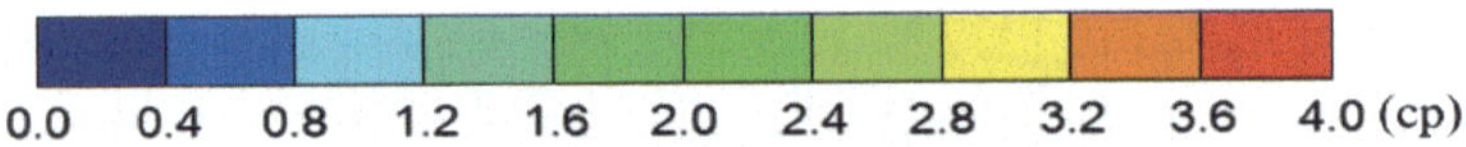

Fig. 3.4 (continued)

separated into three years (2007–2010) for water flooding and ten years (2010–2020) for continuous CO_2 injection (CGI) or WAG with 1:1 ratio. The initial reservoir and operating conditions are indicated in Table 3.8.

The performance of WAG process is compared to that of water flooding and CGI for the base model without asphaltene deposition. As injected CO_2 acts as a solvent in the reservoir, it reduces oil viscosity as shown in Fig. 3.6. Since oil viscosity is not affected by compositional change during water flooding, viscosity does not change much and retains initial value around 3 cp. In contrast, CO_2 injection during WAG decreases oil viscosity to 1 cp, resulting in improved displacement efficiency. After oil is displaced, the viscosity starts to increase from blue area to injector because intermediate components in residual oil are displaced, and the residual oil has much heavier components than initial state. Oil density rises during water flooding for three years from 2007 to 2010 due to pressure increment, and it starts to decrease after CO_2 contact in CGI and WAG processes. It increases again because of heavy components in residual oil as indicated in Fig. 3.7. Having greater ability of pressure maintenance by water injection, WAG has highest oil recovery among the processes (Figs. 3.8 and 3.9). WAG has 17% higher recovery than water flooding and 9% than CGI. At the first stage of CO_2 injection in CGI, oil recovery is lower than water flooding because of lower mobility. However, it soon overtakes as 8% higher than that of water flooding.

Based on the WAG base model, CO_2 and CO_2-LPG models have been developed with formation damages by asphaltene deposition. Deposited asphaltenes cause not only the reduction of absolute permeability and porosity but also wettability change. It is assumed that asphaltene deposition is the only factor to affect relative permeability and contact angle changes, so the last bracket in Eqs. 2.65 and 2.66 in Chap. 2 are

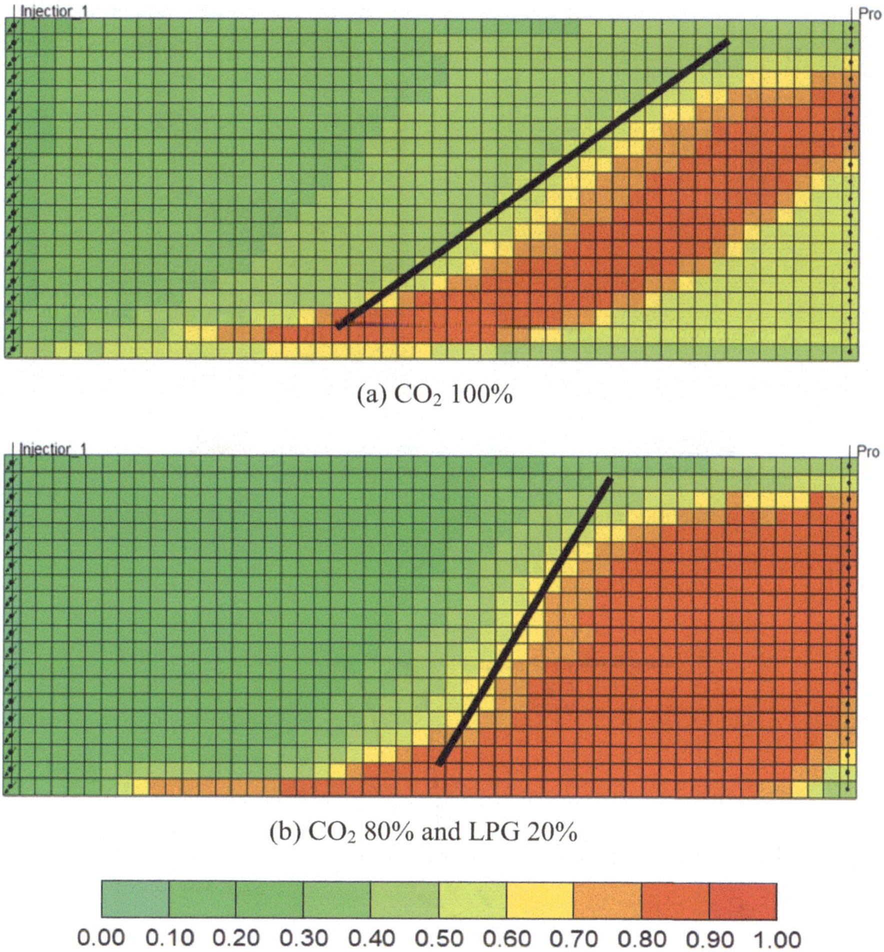

(a) CO_2 100%

(b) CO_2 80% and LPG 20%

Fig. 3.5 Oil saturation distribution with **a** CO_2 100% and **b** CO_2 80% + LPG 20% at 0.6 PV injection

Table 3.8 Initial and operating conditions for WAG process

Parameters	Values
Initial pressure (psi)	4,000
Temperature (°F)	212
Permeability (md)	100
Porosity (%)	25
Initial oil saturation (S_o)	0.7
Connate water saturation (S_w)	0.3
Producing bottom hole pressure (psi)	2,300
Total injection of CO_2 (PV)	1

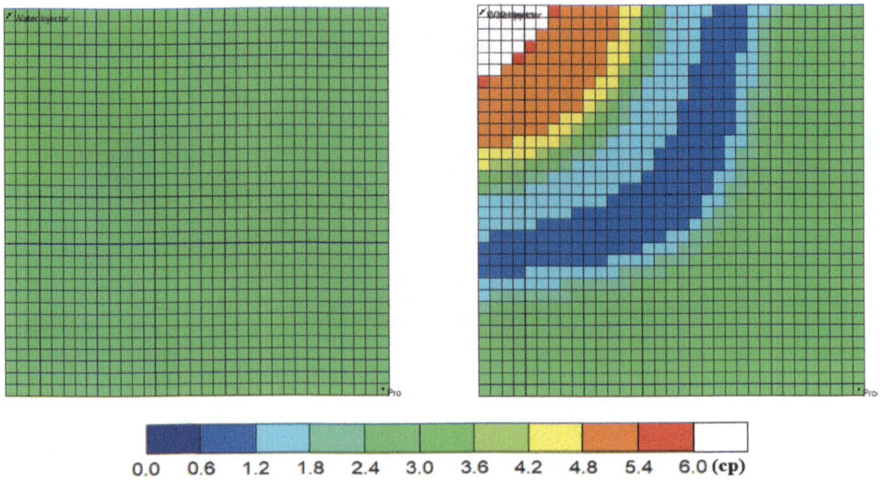

(a) Waterflooding

(b) WAG process

Fig. 3.6 Oil viscosity during **a** water flooding and **b** WAG process after 1-year injection

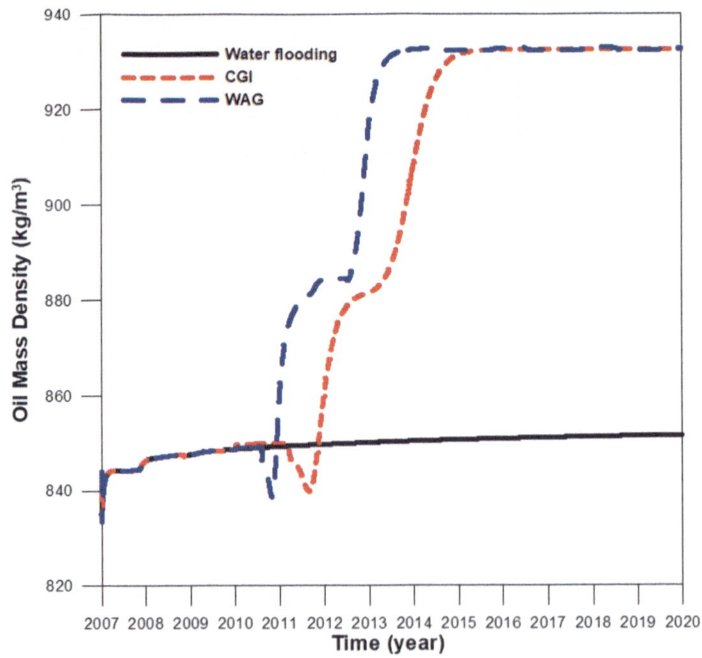

Fig. 3.7 Oil mass density at the middle of reservoir during water flooding, CGI, and WAG processes

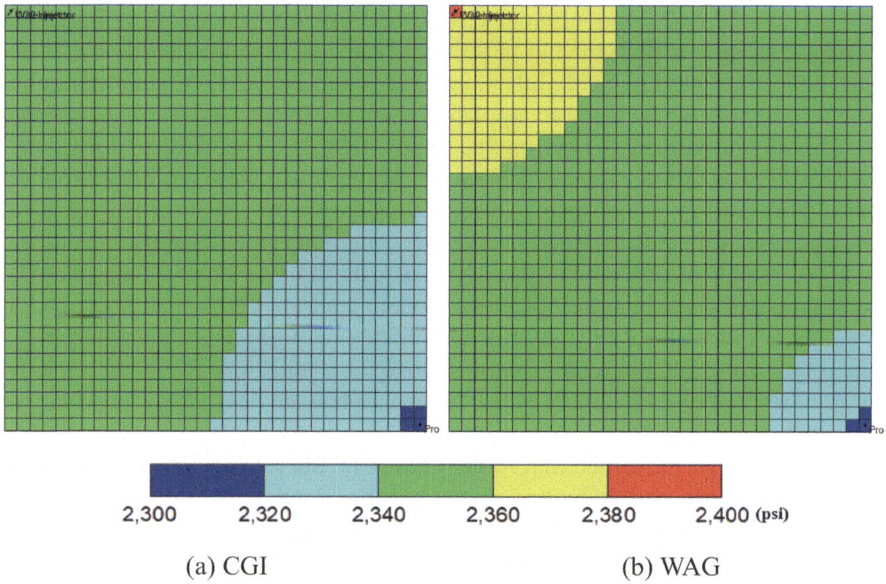

Fig. 3.8 Pressure during **a** CGI and **b** WAG processes after 1-cycle of injection

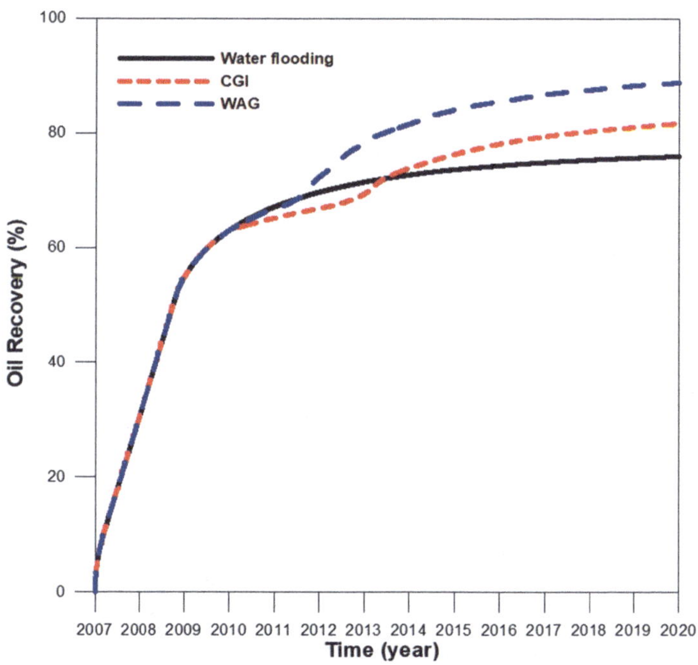

Fig. 3.9 Oil recovery during water flooding, CGI, and WAG processes

equal to one because T_w and N_T are not associated with residual trapping. The initial reservoir condition is presumed to be strongly water-wet condition with 0° contact angle, and it is altered to be slightly water-wet condition with 80° after maximum asphaltene deposition by using interpolation method. The data of reference relative permeability and Burke Oil 1 is from the same reference for reliability (Burke et al. 1990; Fazelipour 2011). Since relative permeability from the reference is slightly water-wet condition, initial data is inversely estimated by equation from 2.64 to 2.66. The reservoir alters to slightly water-wet condition gradually depending on the amount of asphaltene deposition (Fig. 3.10). Also, the input parameters for asphaltene flocculation and deposition in asphaltene modeling part of Chap. 2 are indicated in Table 3.9. The data of input parameters of asphaltene deposition is obtained from the same reference of oil and relative permeability for a reliability (Fazelipour 2011). It is assumed that re-entrainment of flocculated asphaltene does not occur with zero value of β.

In this book, it is assumed that re-dissolution of flocculated asphaltene and re-entrainment of deposited asphaltene do not occur as the data from reference, so the amount of asphaltene deposition does not decrease through the processes. Asphaltene deposition occurs during WAG process not only near producer but also everywhere in the reservoir as shown in Fig. 3.11. The amount of asphaltene deposition increases from near producer to injector because of pressure gradient, so less asphaltene deposits near injector than other regions due to pressure maintenance by water and

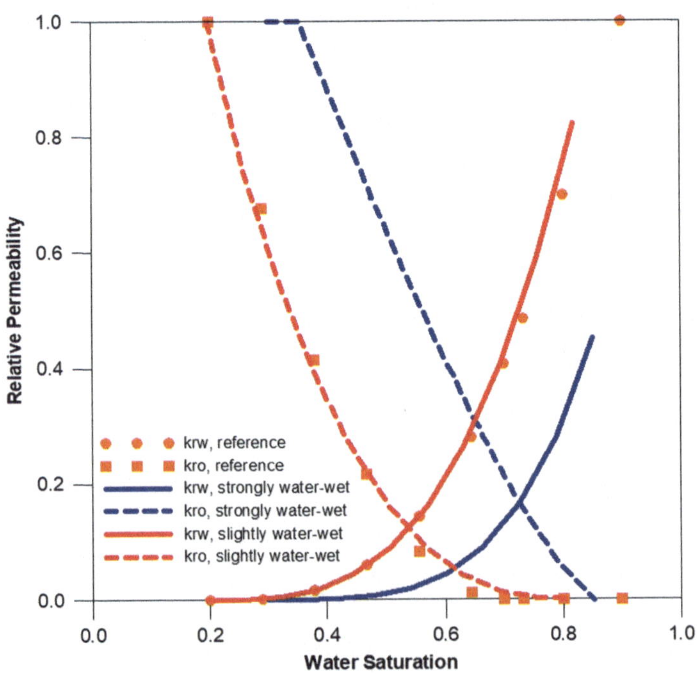

Fig. 3.10 Water and oil relative permeability before and after asphaltene deposition matched to reference data

Table 3.9 Input parameters for asphaltene flocculation and deposition of Burke Oil 1

Parameters	Values
K_{12}	0.1
K_{21}	0.08
α	0.01
β	0
$v_{cr,o}$	0
γ	0.05
σ	150

gas injection. Asphaltene deposition is the function of pressure and oil composition change during WAG, so the deposition occurs in two stages. Asphaltene deposits due to only pressure reduction until 2010, and it has increased near injector since 2010 because of the gas injection (Fig. 3.11a). The second stage of deposition increment at center of reservoir and near producer follows in 2011 and 2012, which is delayed for one- and two-year compared with near injector region due to the travel time from injector to producer (Fig. 3.11). Although LPG improves the displacement efficiency by reducing MMP, it also causes more severe asphaltene problems due to swelling effect. 10% LPG addition model is compared to CO_2 WAG model to investigate the effect of LPG addition on formation damage by asphaltene deposition. As indicated in Fig. 3.11, LPG addition enhances the amount of asphaltene deposition after oil contacts with injected gas everywhere in the reservoir (2010 near the injector, 2011 at the middle, and 2012 near the producer). The formation damages are revealed as oil resistance factor and relative permeability changes. Figure 3.12 shows the oil resistance factor which is the ratio of initial porosity and permeability and that after the formation damage. The oil resistance factor in LPG addition model is 4% greater than that in CO_2 WAG. Also, the damage of wettability change can be interpreted by relative permeability change. Water relative permeability does not change and only follows input data of green line in Fig. 3.13 when asphaltene is not considered because formation damages do not occur. However, it is enhanced at the same water saturation as a black line after wettability alteration during CO_2 WAG because water relative permeability is increased and oil relative permeability is reduced after asphaltene deposition as indicated in Fig. 3.10 for relative permeability curves. When CO_2-LPG WAG is conducted, the increment is more significant as a red line showing 70% greater increment of water relative permeability than that of CO_2 WAG without asphaltene at 0.6 water saturation because of the increase of asphaltene deposition amount. Finally, the result of oil recovery with the models is shown in Fig. 3.14. When the asphaltene models are not considered, LPG enhances the WAG performance by enhanced displacement efficiency with the effect of miscibility improvement. However, these can be overestimated if formation damages with asphaltene deposition are not reflected (Fig. 3.14).

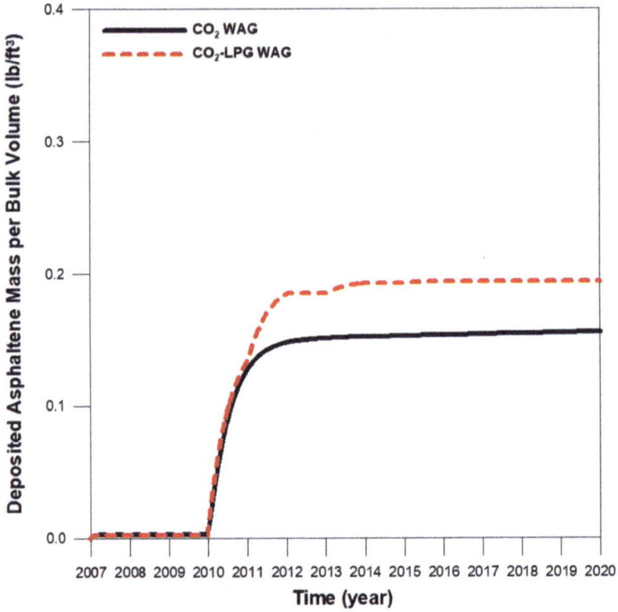

(a) Near injector

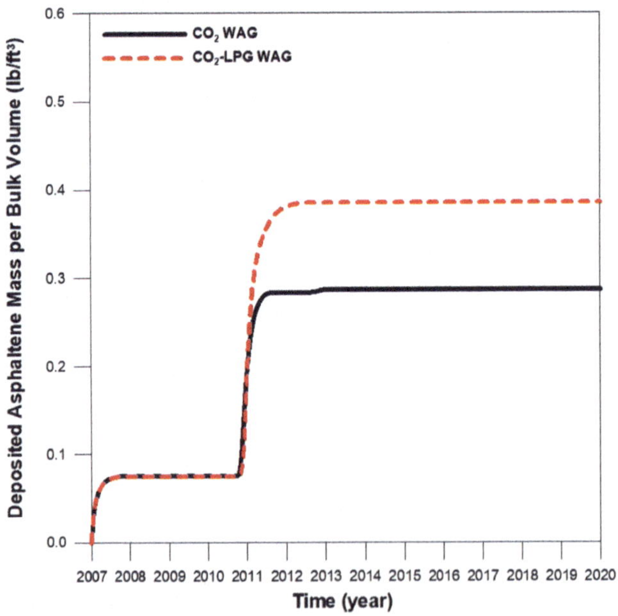

(b) Center of reservoir

Fig. 3.11 The amount of asphaltene deposition per bulk volume **a** near injector, **b** center of reservoir, and **c** near producer during CO_2 and CO_2-LPG WAG

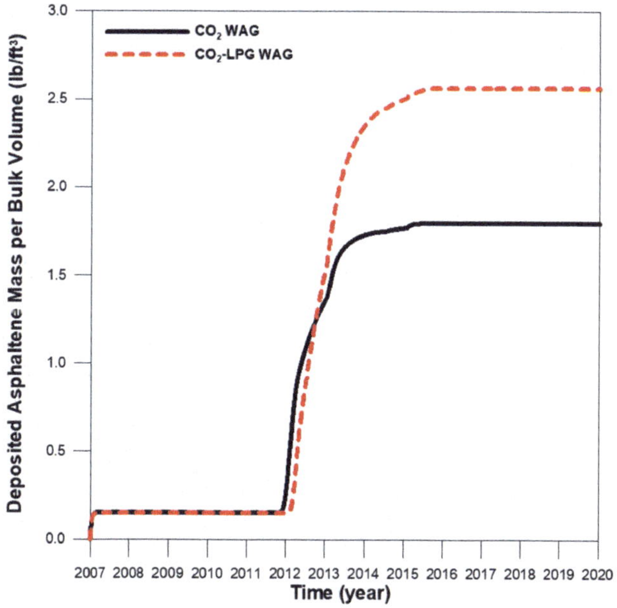

(c) Near producer

Fig. 3.11 (continued)

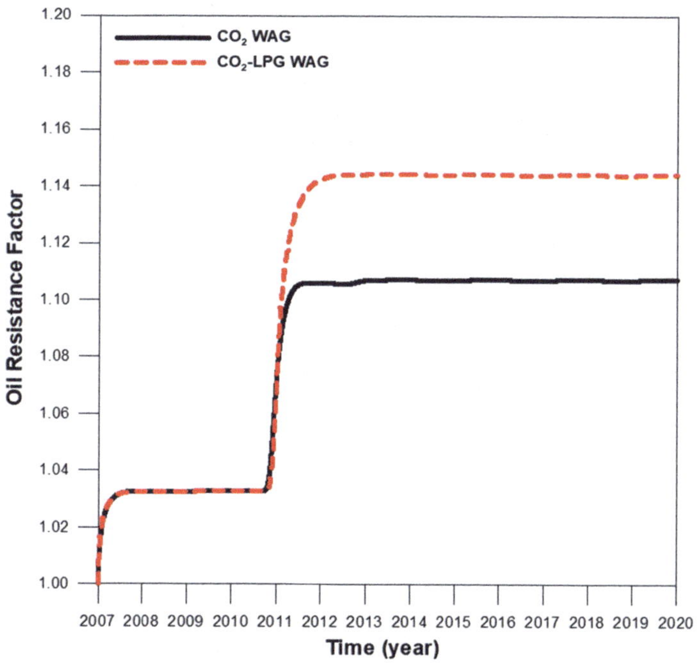

Fig. 3.12 Oil resistance factor at the center of reservoir during CO_2 WAG and CO_2-LPG WAG

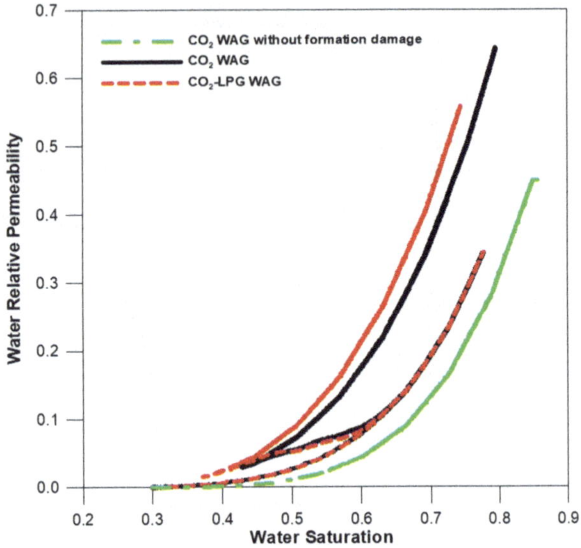

Fig. 3.13 Water relative permeability with wettability alteration during CO_2 WAG and CO_2-LPG WAG

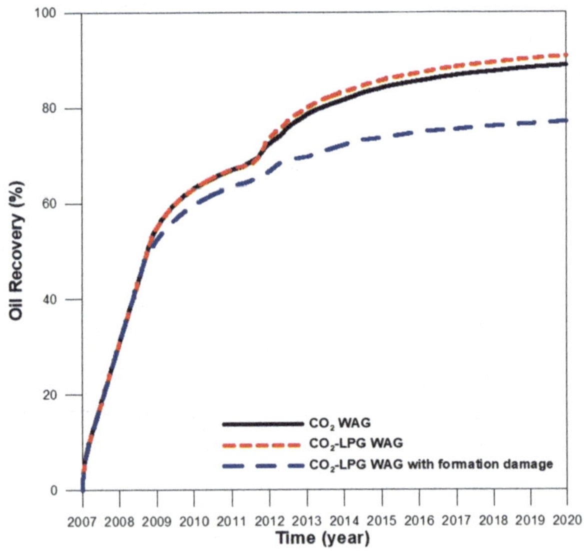

Fig. 3.14 Oil recovery with and without asphaltene deposition during CO_2 and CO_2-LPG WAG processes

References

Abedini A, Mosavat N, Torabi F (2014) Determination of minimum miscibility pressure of crude oil–CO_2 system by oil swelling/extraction test. Energy Technology 2(5):431–439. https://doi.org/10.1002/ente.201400005

Alavian SA, Whitson CH, Martinsen SO (2014) Global component lumping for EOS calculations. In: Paper presented at the SPE annual technical conference and exhibition, Amsterdam, The Netherlands, 27–29 Oct 2014. https://doi.org/10.2118/170912-ms

Alston RB, Kokolis GP, James CF (1985) CO_2 minimum miscibility pressure: a correlation for impure CO_2 streams and live oil systems. SPE Journal 25(02):268–274. https://doi.org/10.2118/11959-PA

Brown JS, Al Kobaisi MS, Kazemi H (2013) Compositional phase trapping in CO_2 WAG simulation. In: Paper presented at the SPE reservoir characterization and simulation conference and exhibition, Abu Dhabi, UAE, 16–18 Sept 2013. https://doi.org/10.2118/165983-ms

Burke NE, Hobbs RE, Kashou SF (1990) Measurement and modeling of asphaltene precipitation. J Petrol Technol 42(11):1440–1446. https://doi.org/10.2118/18273-PA

Cho J, Kim TH, Chang N, Lee KS (2019) Effects of asphaltene deposition-derived formation damage on three-phase hysteretic models for prediction of coupled CO_2 enhanced oil recovery and storage performance. J Petrol Sci Eng 172:988–997. https://doi.org/10.1016/j.petrol.2018.09.006

Cho J, Kim TH, Lee KS (2016) Compositional modeling of hybrid CO_2 EOR with intermediate hydrocarbon solvents to analyze the effect of toluene on asphaltene deposition. J Petrol Sci Eng 146:940–948. https://doi.org/10.1016/j.petrol.2016.08.005

Cho J, Lee KS (2016) Effects of hydrocarbon solvents on simultaneous improvement in displacement and sweep efficiencies during CO_2-enhanced oil recovery. Pet Sci Technol 34(4):359–365. https://doi.org/10.1080/10916466.2015.1137940

Chugh S, Baker R, Cooper L, Spence S (2000) Simulation of horizontal wells to mitigate miscible solvent gravity override in the Virginia hills margin. J Can Pet Technol 39(02):28–34. https://doi.org/10.2118/00-02-02

Coats KH (1985) Simulation of gas condensate reservoir performance. J Petrol Technol 37(10):1870–1886. https://doi.org/10.2118/10512-PA

Darabi H, Sepehrnoori K, Kalaei MH (2012) Modeling of wettability alteration due to asphaltene deposition in oil reservoirs. In: Paper presented at the SPE annual technical conference and exhibition, San Antonio, Texas, USA, 8–10 Oct 2012. https://doi.org/10.2118/159554-ms

Darabi H, Shirdel M, Kalaei MH, Sepehrnoori K (2014) Aspects of modeling asphaltene deposition in a compositional coupled wellbore/reservoir simulator. In: Paper presented at the SPE improved oil recovery symposium, Tulsa, Oklahoma, USA, 12–16 April 2014. https://doi.org/10.2118/169121-ms

Dorshow RB (1995) The simultaneous measurement of interfacial tension and oil viscosity at reservoir conditions for Prudhoe bay fluids by surface laser light scattering spectroscopy. SPE Adv Technol Ser 3(01):120–128. https://doi.org/10.2118/22633-PA

Elsharkawy AM, Poettmann FH, Christiansen RL (1992) Measuring minimum miscibility pressure: slim-tube or rising-bubble method? In: Paper presented at the SPE/DOE enhanced oil recovery symposium, Tulsa, Oklahoma, 22–24 April 1992. https://doi.org/10.2118/24114-ms

Fazelipour W (2011) Predicting asphaltene precipitation in oilfields via the technology of compositional reservoir simulation. In: Paper presented at the SPE international symposium on oilfield chemistry, Woodlands, Texas, USA, 11–13 April 2011. https://doi.org/10.2118/141148-ms

Gonzalez K, Barrufet MA, Nasrabadi H (2014) Development of a compositional reservoir simulator including asphaltene precipitation from a thermodynamic consistent model. In: Paper presented at the SPE Latin America and Caribbean petroleum engineering conference, Maracaibo, Venezuela, 21–23 May 2014. https://doi.org/10.2118/169401-ms

Huang SS, Dyer SB (1992) Miscible displacement in the Weyburn reservoir a laboratory study. In: Paper presented at the technical meeting/petroleum conference of the South Saskatchewan section, Regina, 7–9 Oct 1992. https://doi.org/10.2118/ss-92-20

Jessen K, Orr FM (2008) On interfacial tension measurements to estimate minimum miscibility pressures. SPE Reserv Eval Eng 11(05):933–939. https://doi.org/10.2118/110725-PA

Joergensen M, Stenby EH (1995) Optimization of pseudo-component selection for compositional studies of reservoir fluids. In: Paper presented at the SPE annual technical conference and exhibition, Dallas, Texas, 22–25 Oct 1995. https://doi.org/10.2118/30789-ms

Jun W, Feng W, Deping Z, Guojun Y, Ruosheng P, Shuai X (2013) CO_2 flooding WAG safety control technology. In: Paper presented at the SPE Asia Pacific oil and gas conference and exhibition, Jakarta, Indonesia, 22–24 Oct 2013. https://doi.org/10.2118/165751-ms

Kumar N, Von Gonten WD (1973) An investigation of oil recovery by injecting CO_2 and LPG mixtures. In: Paper presented at the fall meeting of the society of petroleum engineers of AIME, Las Vegas, Nevada, 30 Sept 1973. https://doi.org/10.2118/4581-ms

Lake LW (2010) Enhanced oil recovery. Society of Petroleum Engineers, Richardson, Texas

Lee ST, Jacoby RH, Chen WH, Culham WE (1981) Experimental and theoretical studies on the fluid properties required for simulation of thermal processes. SPE J 21(05):535–550. https://doi.org/10.2118/8393-PA

Li D, Kumar K, Mohanty KK (2003) Compositional simulation of WAG processes for a viscous oil. In: Paper presented at the SPE annual technical conference and exhibition, Denver, Colorado, 5–8 Oct 2003. https://doi.org/10.2118/84074-ms

Li YK, Nghiem LX, Siu A (1985) Phase behaviour computations for reservoir fluids: effect of pseudo-components on phase diagrams and simulation results. J Can Pet Technol 24(06):29–36. https://doi.org/10.2118/85-06-02

Metcalfe RS (1982) Effects of impurities on minimum miscibility pressures and minimum enrichment levels for CO_2 and rich-gas displacements. SPE J 22(02):219–225. https://doi.org/10.2118/9230-PA

Orr FM (2007) Theory of gas injection processes. Tie-Line Publications, Copenhagen, Denmark

Rao DN (1997) A new technique of vanishing interfacial tension for miscibility determination. Fluid Phase Equilib 139(1):311–324. https://doi.org/10.1016/S0378-3812(97)00180-5

Reid RC, Reid RD, Prausnitz JM, Sherwood TK (1977) The properties of gases and liquids. McGraw-Hill, New York

Righi EF, Royo J, Gentil P, Castelo R, Del Monte A, Bosco S (2004) Experimental study of tertiary immiscible WAG injection. In: Paper presented at the SPE/DOE symposium on improved oil recovery, Tulsa, Oklahoma, 17–21 April 2004. https://doi.org/10.2118/89360-ms

Sayegh S, Huang S, Zhang YP, Lavoie R (2007) Effect of H_2S and pressure depletion on the CO_2 MMP of Zama oils. J Can Pet Technol 46(08):34–41. https://doi.org/10.2118/07-08-03

Sharbatian A, Abedini A, Qi Z, Sinton D (2018) Full characterization of CO_2–oil properties on-chip: solubility, diffusivity, extraction pressure, miscibility, and contact angle. Anal Chem 90(4):2461–2467. https://doi.org/10.1021/acs.analchem.7b05358

Sifuentes WF, Giddins MA, Blunt MJ (2009) Modeling CO_2 storage in aquifers: assessing the key contributors to uncertainty. In: Paper presented at the offshore Europe, Aberdeen, UK, 8–11 Sept 2009. https://doi.org/10.2118/123582-ms

Srivastava RK, Huang SS, Dong M (1999) Asphaltene deposition during CO_2 flooding. SPE Prod Facil 14(04):235–245. https://doi.org/10.2118/59092-PA

Srivastava RK, Huang SS, Dong M (2000) Laboratory investigation of Weyburn CO_2 miscible flooding. J Can Pet Technol 39(02):41–51. https://doi.org/10.2118/00-02-04

Stalkup FI Jr (1983) Status of miscible displacement. J Petrol Technol 35(04):815–826. https://doi.org/10.2118/9992-PA

Talbi K, Maini BB (2008) Experimental investigation of CO_2-based VAPEX for recovery of heavy oils and bitumen. J Can Pet Technol 47(04):1–8. https://doi.org/10.2118/08-04-29

Teklu TW, Alharthy N, Kazemi H, Yin X, Graves RM (2014) Hydrocarbon and non-hydrocarbon gas miscibility with light oil in shale reservoirs. In: Paper presented at the SPE improved oil recovery symposium, Tulsa, Oklahoma, USA, 12–16 April 2014. https://doi.org/10.2118/169123-ms

Teletzke GF, Patel PD, Chen A (2005) Methodology for miscible gas injection EOR screening. In: Paper presented at the SPE international improved oil recovery conference in Asia Pacific, Kuala Lumpur, Malaysia, 5–8 Dec 2005. https://doi.org/10.2118/97650-ms

Whitson CH (1983) Characterizing hydrocarbon plus fractions. SPE J 23(04):683–694. https://doi.org/10.2118/12233-PA

Whitson CH, Michelsen ML (1989) The negative flash. Fluid Phase Equilib 53:51–71. https://doi.org/10.1016/0378-3812(89)80072-X

Yellig WF, Metcalfe RS (1980) Determination and prediction of CO_2 minimum miscibility pressures. J Petrol Technol 32(01):160–168. https://doi.org/10.2118/7477-PA

Zhang K, Jia N, Zeng F (2018) Application of predicted bubble-rising velocities for estimating the minimum miscibility pressures of the light crude oil–CO_2 systems with the rising bubble apparatus. Fuel 220:412–419. https://doi.org/10.1016/j.fuel.2018.01.100

Zhang PY, Huang S, Sayegh S, Zhou XL (2004) Effect of CO_2 impurities on gas-injection EOR processes. In: Paper presented at the SPE/DOE symposium on improved oil recovery, Tulsa, Oklahoma, 17–21 April 2004. https://doi.org/10.2118/89477-ms

Chapter 4
Numerical Modeling of CCS-EOR

Abstract The overall objective of this chapter is to investigate the combined effects of asphaltene and CCS mechanisms on CCS-EOR performance. An integrated model for CCS-EOR with CO_2 and LPG addition considering formation damages by asphaltene deposition and three-phase hysteresis have been described. For CCS-EOR in only CO_2 fluid system, hysteresis effect reflects the CO_2 residual trapping that occupies the pore spaces and displaces water and oil. Therefore, CCS-EOR performance can be underestimated if three-phase effect is not considered. The effects of asphaltene and three-phase hysteresis on CCS and EOR are also explained. If both three-phase hysteresis and asphaltene deposition are applied, CCS-EOR performance is changed because deposited asphaltene improves water mobility and reduces oil and gas mobility. Residual and solubility trappings can be under and overestimated for CCS, respectively, and oil recovery is overestimated for EOR without the integrative model. An integrated model for CO_2-LPG WAG is developed to investigate the interacted mechanisms and the effects of LPG on CCS-EOR performance. Since LPG accelerates the asphaltene deposition, the amount of residual trapping is greater than CO_2 WAG, but the solubility trapping performance is declined in spite of higher reservoir pressure due to more formation damage. Because the enhancement of residual trapping is greater than that of solubility trapping, it can be confirmed that LPG enhances both CCS and EOR performances during coupled CCS-EOR process. However, the utility of LPG application is also affected by CO_2 and LPG prices, so the economic analysis should be accompanied for field practical use.

Keywords Hysteresis · Residual trapping · Solubility trapping · CCS-EOR · Asphaltene deposition

4.1 Effects of Asphaltene on CCS-EOR Mechanisms

CO_2 EOR, which is performed by oil field operator, has a limitation of considering environmental issues like climate change, while CCS, which is usually conducted by the government, is limited due to lack of business model. Therefore, CCS-EOR

K. S. Lee et al., *CO_2 Storage Coupled with Enhanced Oil Recovery*,
https://doi.org/10.1007/978-3-030-41901-1_4

technology for the partnership between them can carry out long-term benefits for both (Ettehadtavakkol et al. 2014). The advantages of improving energy security, tax credits, and EOR benefits increase considerable attention of CCS-EOR (Stevens et al. 2001; Bachu and Shaw 2003; Moritis 2003; Beecy and Kuuskraa 2005; Kheshgi et al. 2010; Agarwal and Parsons 2011; Kuuskraa et al. 2011; Veld et al. 2013). The objectives of CO_2 EOR and storage are incompatible, and usually EOR is a priority because of the economic benefits (Kovscek and Cakici 2005; Leach et al. 2011). To optimize the design parameters considering economic feasibility, Ettehadtavakkol et al. (2014) performed the simulation studies with various WAG ratios. WAG injection process is also applied in this book to enhance both the displacement and sweep efficiencies for EOR and hysteresis and solubility trapping for CCS simultaneously. During WAG injection, asphaltene precipitation occurs in the reservoir (Khanifar et al. 2011). Many researchers have studied the effects of pressure, temperature, and oil composition changes on asphaltene precipitation and deposition and its influence on EOR performance (Correra et al. 2010; Cao and Gu 2013; Darabi et al. 2014; Zanganeh et al. 2015; Wang et al. 2017). However, the effects on CCS mechanisms are available only from Cho et al. (2019). If CCS is applied into an aquifer, we need to consider the two-phase system of water and gas. For CCS-EOR, it is necessary to take account of four-phase system of water, gas, oil, and solid.

To reflect the effects of asphaltene deposition on CCS-EOR, the reservoir simulation models have been developed by including CCS mechanisms from the models in Chap. 3. Firstly, CCS-EOR models with three-phase hysteresis, but not considering formation damage, is developed to investigate only the effect of the hysteresis on the performance. Then, the integrated model with the hysteresis and formation damage by asphaltene deposition is developed. Solubility model is included in all the models. The integrated process and methodologies are indicated in flow chart, Fig. 4.1.

If three-phase hysteresis and formation damage are not considered, the water relative permeability is fixed and follows the initial input curve during both drainage and imbibition processes as the black solid line in Fig. 4.2a. If three-phase hysteresis is considered as the red line, it has its own scanning curve. First drainage curve (decreasing water saturation) does not trail the initial input water relative permeability, and the next imbibition curve (increasing water saturation) has also its own relative permeability that means irreversibility. While the imbibition and drainage processes are conducted during WAG, because water and gas are trapped, water mobility is reduced causing the lower water relative permeability. When the rock wettability is altered from strongly to slightly water wet condition by asphaltene deposition, the wetting phase (water) mobility is enhanced. Because the attraction between the rock surface and water becomes weak, the scanning curves are developed above the models without formation damage as the blue line in Fig. 4.2a. Unlike the red line, the scanning curve with asphaltene deposition shows the aspect of increase in drainage and imbibition processes because water mobility is enhanced gradually. In case of gas, gas relative permeability curve without hysteresis and asphaltene has the same tendency as that of water as indicated in Fig. 4.2b. When only the effect of three-phase hysteresis is reflected, the curves are shifted to the bottom right as the red line that indicates the reduction of gas mobility. As the processes progress, the new cycle

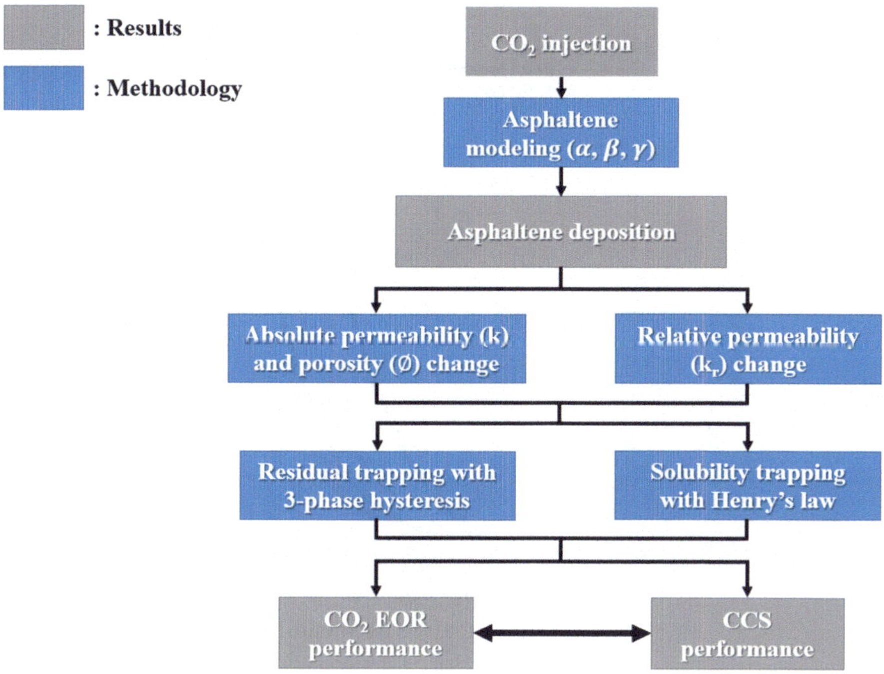

Fig. 4.1 Flowchart of the methodology applied to integrated model for CCS-EOR

of curves is moved to the right continuously. Lower gas mobility with asphaltene model than that without asphaltene deposition causes the gas scanning curve shift to the bottom right as the blue line in Fig. 4.2b. Since it leads to the improvement of hysteresis effect with asphaltene deposition, the amount of residual trapping is 49% larger with formation damage than that without it at the end of process, and the CO_2 trapping does not occur without hysteresis effect (Fig. 4.3). The fluctuation of Fig. 4.3 is caused by imbibition and drainage processes during WAG. When water is injected in imbibition process after gas injection for each cycle of WAG, CO_2 is trapped in the pore by hysteresis effect. In contrast, the trapped gas combines with CO_2 stream which is injected for the drainage after previous water injection, so it departs from residual position in the pores resulting in the decrement of residual trapping performance. For this reason, the amount of residual trapping decreases in even number year due to CO_2 injection but increases in odd number year with water injection.

Figure 4.4 shows the average saturation of gas, water, and oil in the reservoir. The residual gas saturation without hysteresis is the lowest among the models because gas mobility is not reduced and the greatest as indicated in Fig. 4.4a, and most of CO_2 is produced. If hysteresis is considered, the residual gas saturation is increased as the red line due to lower gas mobility and trapped CO_2. It becomes greater with asphaltene deposition as the blue line than other models due to much less gas mobility. As

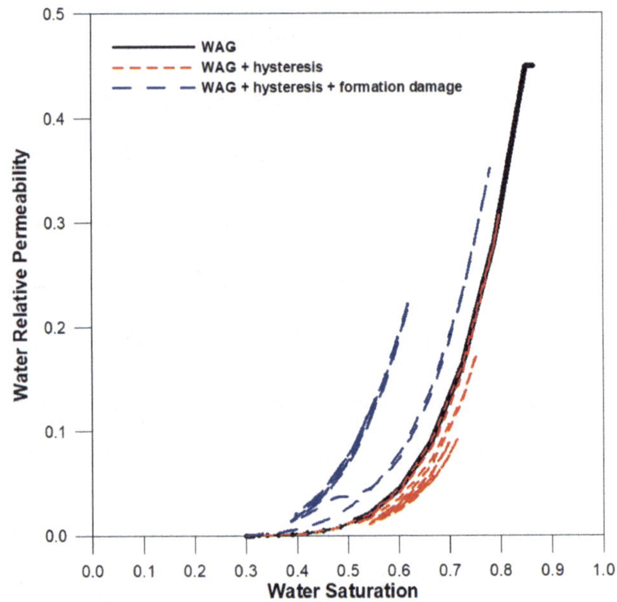

(a) Water relative permeability

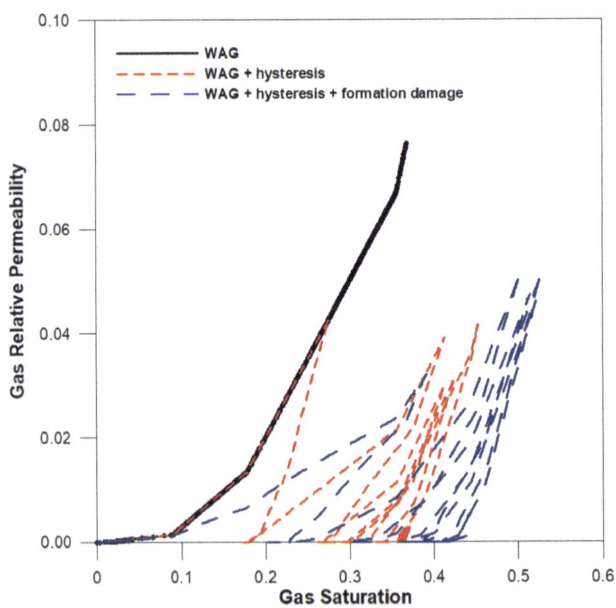

(b) Gas relative permeability

Fig. 4.2 **a** Water and **b** gas relative permeability scanning curves at the center of reservoir for WAG, +three-phase hysteresis, and +formation damage models

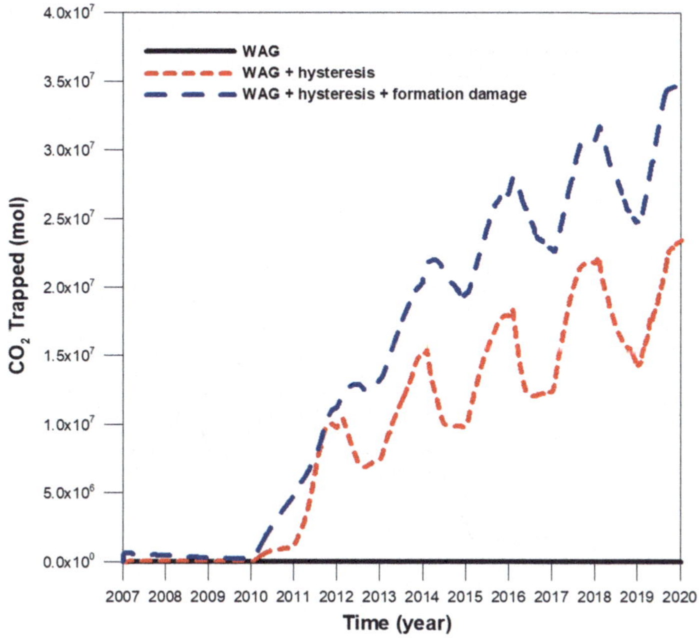

Fig. 4.3 The amount of trapped CO_2 for only WAG, +WAG hysteresis, and +formation damage models

trapped CO_2 occupies the pore spaces, liquids (water and oil) can be displaced as the volume of CO_2 residual trapping. The amount of replaced water and oil is determined by the mobility which is expressed in relative permeability. According to Fig. 4.4b, c, residual water and oil saturations of only hysteresis model are lower than those in WAG model without any effects because both water and oil are displaced by trapped CO_2. In contrast, if asphaltene deposition is considered, residual water saturation is reduced. Because integrated model has the largest gas saturation and water mobility is enhanced, more water is displaced. Most of trapped fluid is oil due to reduced oil mobility after formation damage, and relatively less oil is expelled from the pore space showing the greatest residual oil saturation (Fig. 4.4c). Both WAG only and WAG with hysteresis models have similar amount of dissolved CO_2 in water. The reason is that residual water and gas saturations are lower and higher, respectively, with hysteresis model than that in WAG only case complementing each other as shown in Fig. 4.5. However, the performance of solubility trapping is decreased with formation damage effect due to less remaining water in the reservoir although more gas is residual. The amount of dissolved CO_2 is 7% smaller than that of only WAG model.

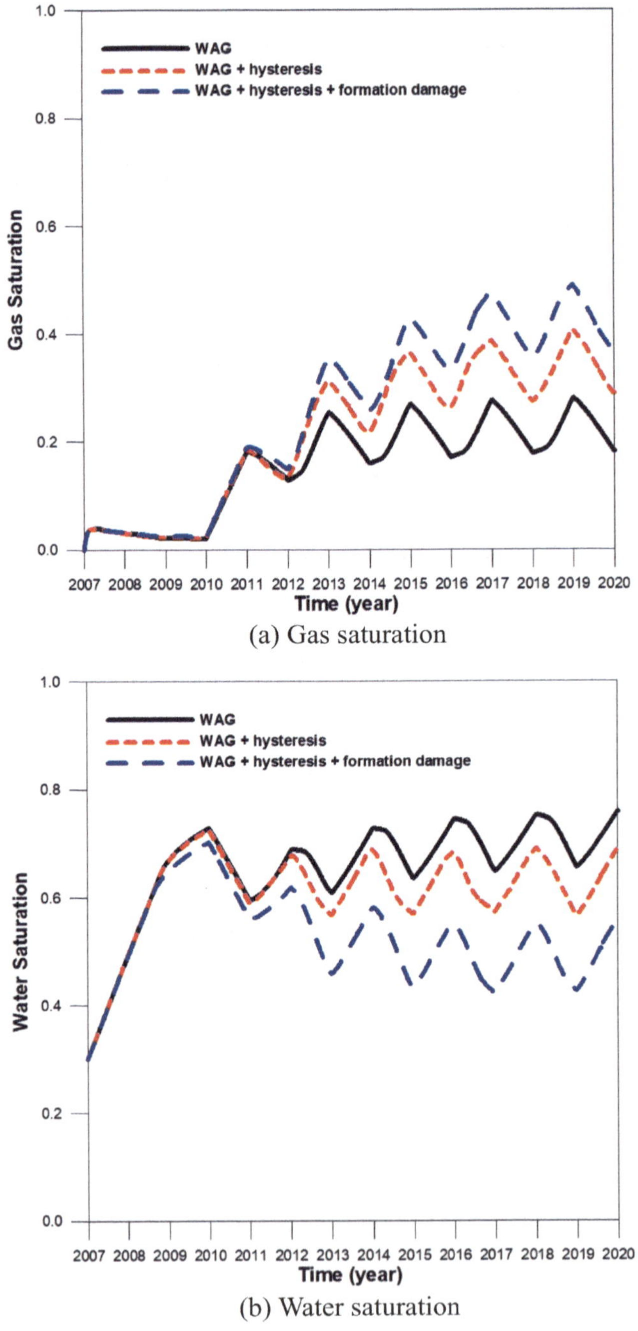

Fig. 4.4 Residual saturation of **a** gas, **b** water, and **c** oil in the reservoir for three different models

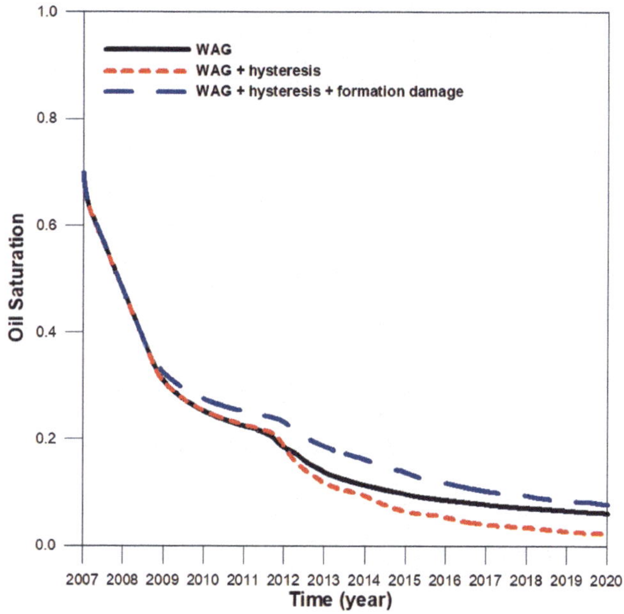

(c) Oil saturation

Fig. 4.4 (continued)

Finally, three-phase hysteresis and formation damage affect the performances of water production and oil recovery. Increased water mobility by formation damage causes half-year earlier water breakthrough and 13 and 22% greater water production than those of hysteresis and WAG only model (Fig. 4.6a). In contrast, oil recovery is 9 and 4% lower than three-phase hysteresis and WAG only models because of declined oil mobility (Fig. 4.6b). The results show that residual and solubility trapping performances can be underestimated and overestimated, respectively, in CCS unless formation damage is considered. Oil recovery, which is affected by CCS performance, is overvalued in EOR during combined CCS-EOR process if formation damage by asphaltene is not reflected.

4.2 Effects of LPG on CCS-EOR Performance

The presence of impurities in the CO_2 stream during CCS in aquifer has an effect on CO_2 storage capacity, injectivity, geochemical reactions, and corrosion of pipeline because of changes in phase behavior and its properties (Wang et al. 2011). The influences of impurities on CCS have been studied including pipeline transport (Bachu and Gunter 2005; Palandri and Kharaka 2005; Seevam et al. 2007; De Visser et al.

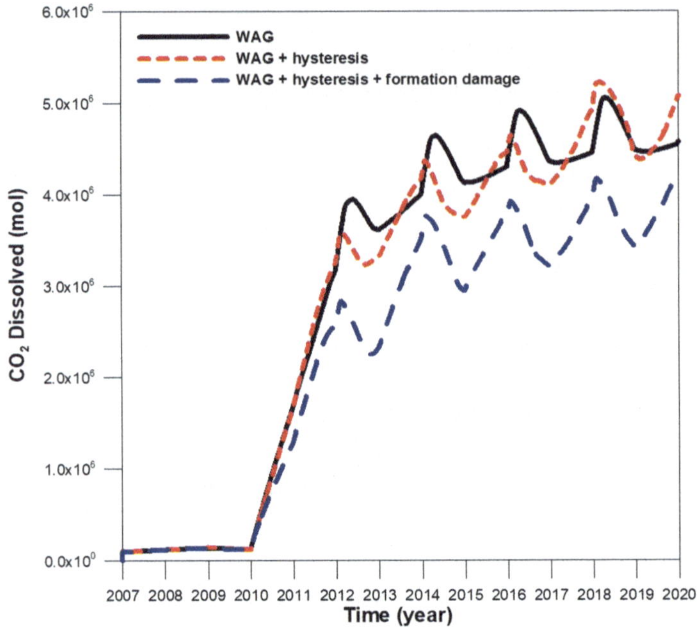

Fig. 4.5 The amount of dissolved CO_2 for only WAG, +WAG hysteresis, and +formation damage models

2008; Oosterkamp and Ramsen 2008; Sass et al. 2009; Ellis et al. 2010; Blanco et al. 2012).

Studies to investigate the effects of impurities on combined CCS-EOR are insufficient and more complicated because of the existence of oil that causes various reactions. Since LPG affects the asphaltene deposition that causes formation damages and CCS mechanisms, the effects should be considered to predict the precise performances of CCS-EOR. To confirm the influence of LPG on both EOR and CCS performances, the reservoir simulation models have been developed from the previous integrated model. If the maximum volume fraction of asphaltene deposition and grid block volumes reaches 0.006 by LPG addition in each grid block, it finally becomes slightly water-wet condition by interpolation method. The LPG mole fraction of 10% is applied in injection gas stream for CO_2-LPG injection, and the injection rate and total amount of injection gas are set as the same of that in CO_2 and CO_2-LPG WAG models.

Figure 4.7 indicates the relative permeability curves of gas and water during the processes. The black solid line in Fig. 4.7a shows the scanning curve during CO_2 WAG with three-phase hysteresis and formation damages by asphaltene deposition, so it is the same as the blue line in Fig. 4.2b. When wettability change is accelerated by

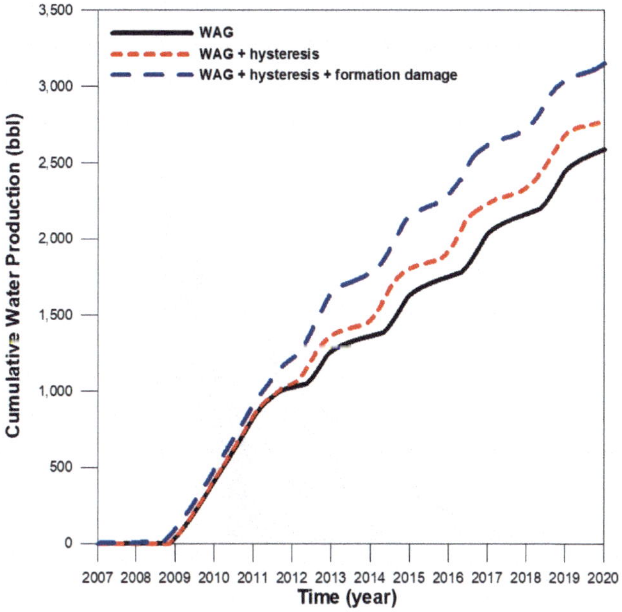

(a) Water production

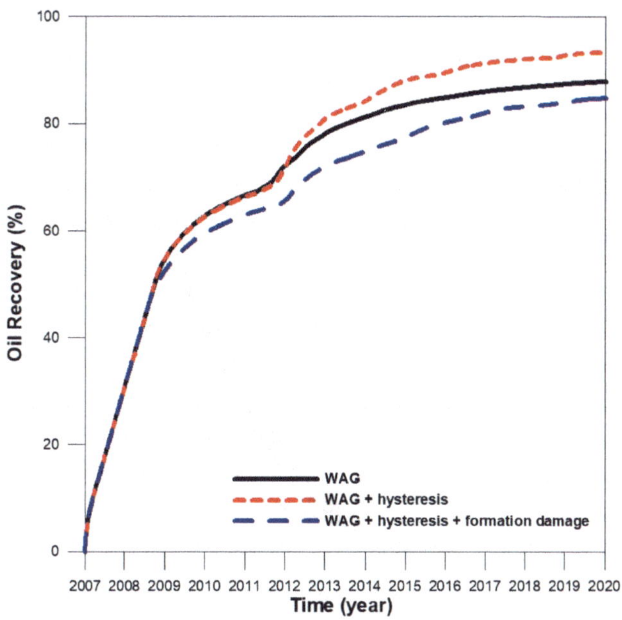

(b) Oil recovery

Fig. 4.6 Cumulative water production and oil recovery for WAG, +WAG hysteresis, and + formation damage models

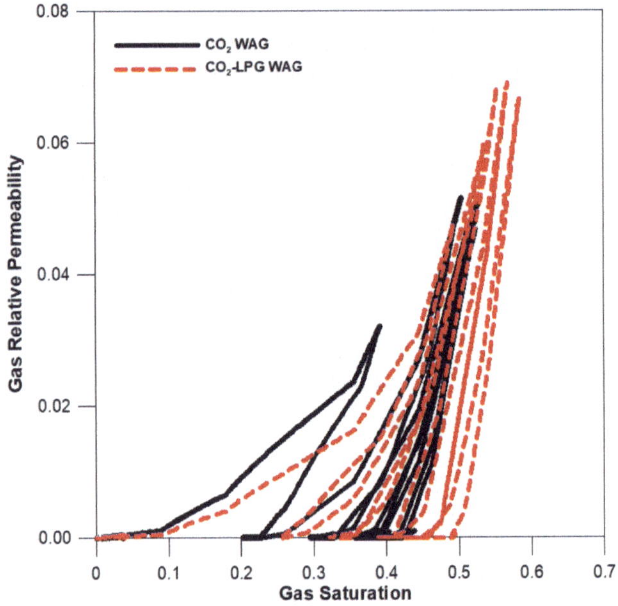

(a) Gas relative permeability

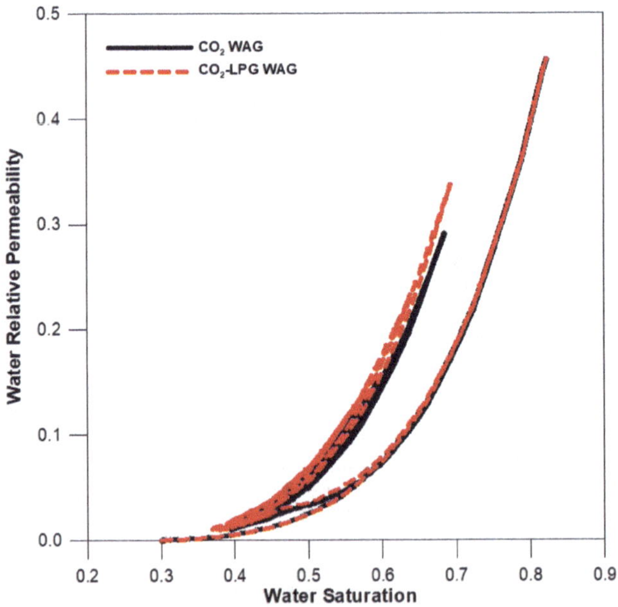

(b) Water relative permeability

Fig. 4.7 **a** Gas and **b** water relative permeability scanning curves at the middle of reservoir for CO_2 and CO_2-LPG WAG

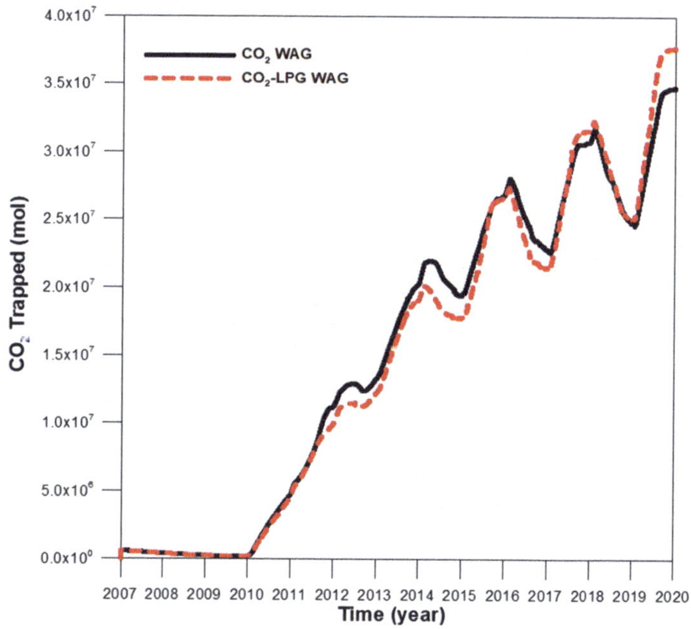

Fig. 4.8 The amount of trapped CO_2 for CO_2 and CO_2-LPG WAG models

LPG addition into CO_2 stream because of increased amount of asphaltene deposition as shown in Fig. 3.11 in Chap. 3, the gas mobility becomes lower than that of CO_2 WAG. Gas relative permeability is shifted to the bottom right having the lower value at the same gas saturation at each WAG cycle. Lower gas mobility increases the residual trapping performance by hysteresis. In spite of less injected CO_2 than that of CO_2 WAG, 9% greater amount of CO_2 is trapped. Though total injection gas is the same, 10% LPG mole fraction takes the portion of injection stream (Fig. 4.8). Contrary to this, LPG enhances the water mobility that has higher water relative permeability as shown in Fig. 4.7b. The improved water mobility and gas hysteresis may cause that increased trapped CO_2 replaces more water, resulting in lower residual water saturation (Fig. 4.9). As a result, the amount of dissolved CO_2 in CO_2-LPG WAG process is 4% lower than CO_2 WAG process (Fig. 4.10). However, it does not directly indicate that LPG reduces the performance of CO_2 solubility trapping. Because the amount of injected CO_2 is different, the ratio of remained CO_2 in the reservoir and injected CO_2 should be compared for two different models (Fig. 4.11). It shows that the remaining CO_2 in the reservoir depends on trapping mechanisms, such as residual trapping, solubility trapping, and movable CO_2 which has the potential to be trapped later. As indicated in Fig. 4.11, residual trapped and movable CO_2 is 4 and

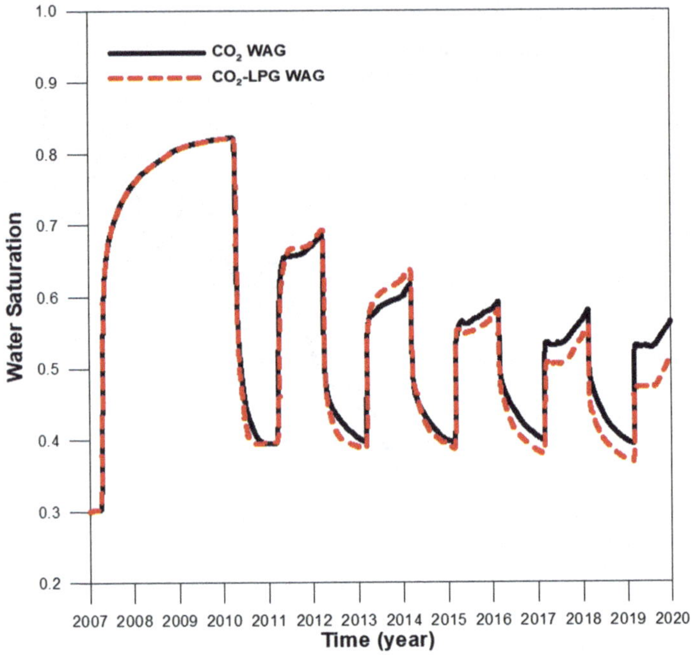

Fig. 4.9 Residual water saturation at the middle of reservoir for CO_2 and CO_2-LPG WAG

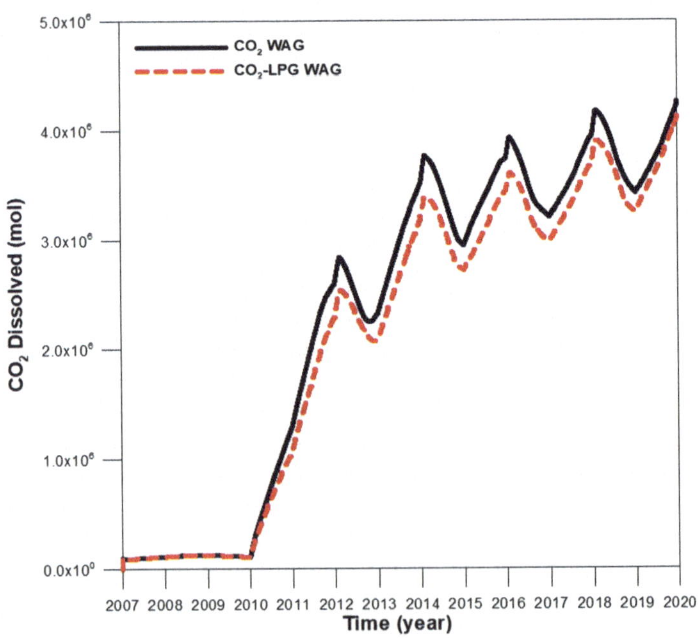

Fig. 4.10 The amount of dissolved CO_2 for CO_2 and CO_2-LPG WAG

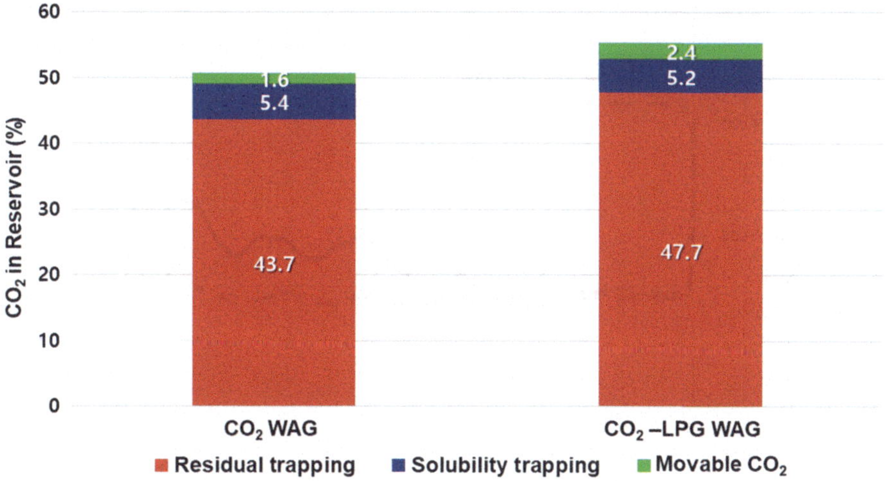

Fig. 4.11 Ratio of residual CO_2 in the reservoir depending on trapping mechanisms (residual trapping, solubility trapping, and movable CO_2)

0.8% higher than CO_2 WAG case and solubility trapping is 0.2% lower. The ratio of stored CO_2 to injected CO_2 is 50.7 and 55.3% for CO_2 WAG and CO_2-LPG WAG, respectively. CCS performances are improved as 9% by LPG addition.

CO_2 solubility trapping is affected by reservoir pressure as well as residual water and CO_2 saturation. As indicated in a blue line of Fig. 4.12, if formation damage is not applied, reservoir pressure does not increase as much as asphaltene models. It means that damage such as pore plugging causes the increment of pressure. As LPG enhances asphaltene deposition, the pressure becomes higher as a red line. Although more formation damage increases average reservoir pressure that usually enhances gas solubility in water, the ratio of dissolved CO_2 in water to injected CO_2 is smaller. However, according to the models of CO_2-LPG WAG with the same CO_2 mol injection as an example, the CO_2 solubility trapping performance is enhanced because of the increased reservoir pressure by LPG addition (Fig. 4.13). It indicates that the solubility trapping performance depends on the mole fraction of CO_2 and LPG. Therefore, the economic analysis should be performed because the price of LPG is much higher than CO_2. Eventually, LPG enhances the oil production due to improved displacement efficiency which is 4% higher than that of CO_2 WAG, so performances of CCS and EOR can be ameliorated simultaneously (Fig. 4.14).

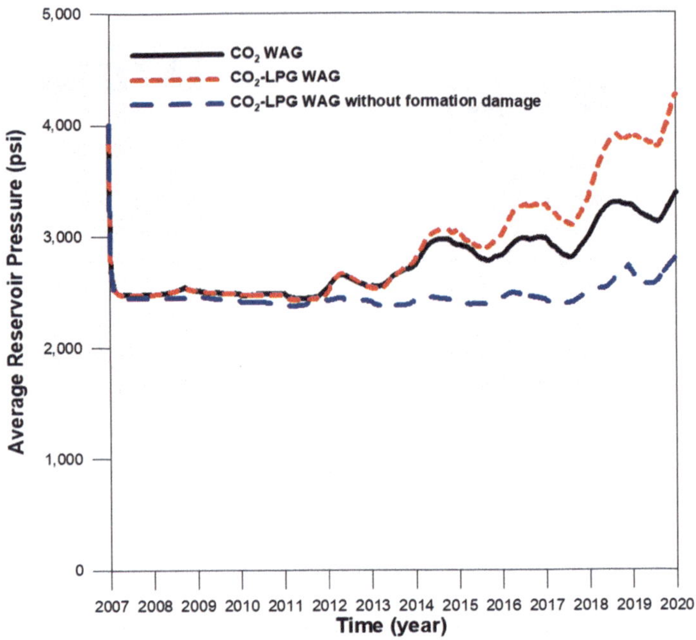

Fig. 4.12 Average reservoir pressure during CO_2 and CO_2-LPG WAG processes

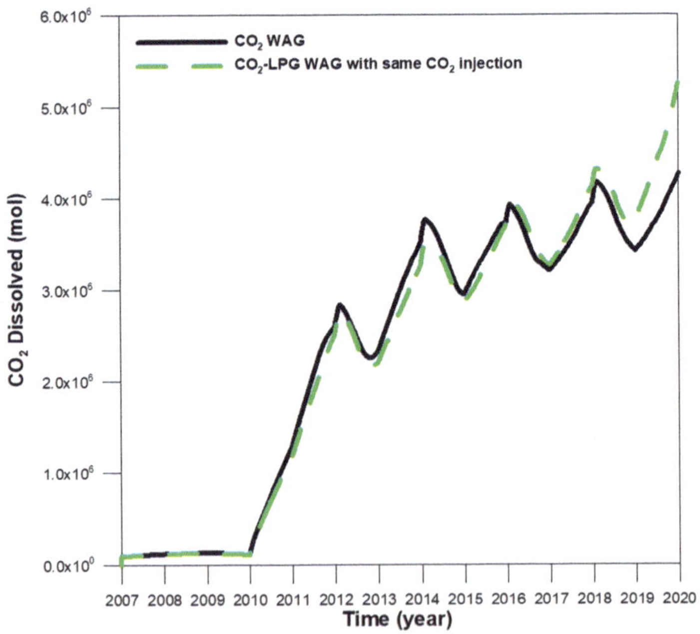

Fig. 4.13 The amount of dissolved CO_2 for CO_2 and CO_2-LPG WAG models with the same amount of CO_2 injection

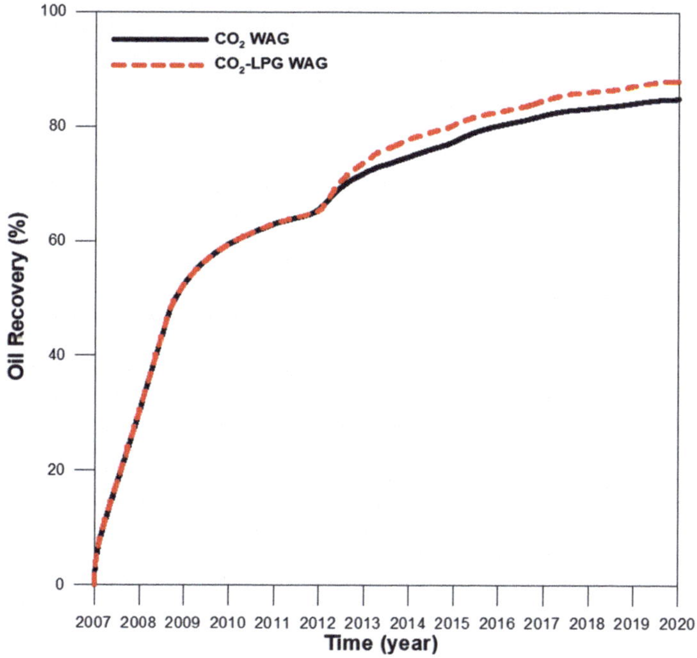

Fig. 4.14 Oil recovery with formation damage and three-phase hysteresis during CO_2 and CO_2-LPG WAG processes

References

Agarwal A, Parsons J (2011) Commercial structures for integrated CCS-EOR projects. Energy Procedia 4:5786–5793. https://doi.org/10.1016/j.egypro.2011.02.575

Bachu S, Gunter WD (2005) Overview of acid-gas injection operations in Western Canada. In: Rubin ES, Keith DW, Gilboy CF, Wilson M, Morris T, Gale J, Thambimuthu K (eds) Greenhouse gas control technologies 7. Elsevier Science Ltd., Oxford, pp 443–448

Bachu S, Shaw J (2003) Evaluation of the CO_2 sequestration capacity in Alberta's oil and gas reservoirs at depletion and the effect of underlying aquifers. J Can Pet Technol 42(9):51–61

Beecy DJ, Kuuskraa VA (2005) Basin strategies for linking CO_2 enhanced oil recovery and storage of CO_2 emissions. In: Rubin ES, Keith DW, Gilboy CF, Wilson M, Morris T, Gale J, Thambimuthu K (eds) Greenhouse gas control technologies 7. Elsevier Science Ltd., Oxford, pp 351–359

Blanco ST, Rivas C, Fernández J, Artal M, Velasco I (2012) Influence of methane in CO_2 transport and storage for CCS technology. Environ Sci Technol 46(23):13016–13023. https://doi.org/10.1021/es3037737

Cao M, Gu Y (2013) Oil recovery mechanisms and asphaltene precipitation phenomenon in immiscible and miscible CO_2 flooding processes. Fuel 109:157–166. https://doi.org/10.1016/j.fuel.2013.01.018

Cho J, Kim TH, Chang N, Lee KS (2019) Effects of asphaltene deposition-derived formation damage on three-phase hysteretic models for prediction of coupled CO_2 enhanced oil recovery and storage performance. J Petrol Sci Eng 172:988–997. https://doi.org/10.1016/j.petrol.2018.09.006

Correra S, De Simoni M, Bartosek M, Colombani G, Diatto P, Maddinelli G, Parracello VP (2010) Assessment of asphaltene deposition risk in an EOR intervention through CO_2 injection. In: Paper presented at the SPE annual technical conference and exhibition, Florence, Italy, 19–22 Sep 2010. https://doi.org/10.2118/135001-ms

Darabi H, Shirdel M, Kalaei MH, Sepehrnoori K (2014) Aspects of modeling asphaltene deposition in a compositional coupled wellbore/reservoir simulator. In: Paper presented at the SPE improved oil recovery symposium, Tulsa, Oklahoma, USA, 12–16 April 2014. https://doi.org/10.2118/169121-ms

De Visser E, Hendriks C, Barrio M, Mølnvik MJ, de Koeijer G, Liljemark S, Le Gallo Y (2008) Dynamis CO_2 quality recommendations. Int J Greenhouse Gas Control 2(4):478–484. https://doi.org/10.1016/j.ijggc.2008.04.006

Ellis BR, Crandell LE, Peters CA (2010) Limitations for brine acidification due to SO_2 co-injection in geologic carbon sequestration. Int J Greenhouse Gas Control 4(3):575–582. https://doi.org/10.1016/j.ijggc.2009.11.006

Ettehadtavakkol A, Lake LW, Bryant SL (2014) CO_2-EOR and storage design optimization. Int J Greenhouse Gas Control 25:79–92. https://doi.org/10.1016/j.ijggc.2014.04.006

Khanifar A, Sheykh Alian S, Demiral B, Darman NB (2011) Study of asphaltene precipitation and deposition phenomenon during WAG application. In: Paper presented at the SPE enhanced oil recovery conference, Kuala Lumpur, Malaysia, 19–21 July 2011. https://doi.org/10.2118/143488-ms

Kheshgi HS, Bhore NA, Hirsch R, Parker ME, Teletzke GF, Thomann H (2010) Perspectives on CCS cost and economics. In: Paper presented at the SPE international conference on CO_2 capture, storage, and utilization, New Orleans, Louisiana, USA, 10–12 Nov 2010. https://doi.org/10.2118/139716-ms

Kovscek AR, Cakici D (2005) Geologic storage of carbon dioxide and enhanced oil recovery. II. Cooptimization of storage and recovery. Energy Convers Manag 46:1941–1956. https://doi.org/10.1016/j.enconman.2004.09.009

Kuuskraa V, Leewen TV, Wallace M (2011) Improving domestic energy security and lowering CO_2 emissions with 'next generation' CO_2-enhanced oil recovery. NETL, USA

Leach A, Mason CF, Veld K (2011) Co-optimization of enhanced oil recovery and carbon sequestration. Resour Energy Econ 33(4):893–912. https://doi.org/10.1016/j.reseneeco.2010.11.002

Moritis G (2003) CO_2 sequestration adds new dimension to oil, gas production. Oil Gas J 101(9):39–44

Oosterkamp A, Ramsen J (2008) State-of-the-art overview of CO_2 pipeline transport with relevance to offshore pipelines. POLYTEC, Norway

Palandri JL, Kharaka YK (2005) Ferric iron-bearing sediments as a mineral trap for CO_2 sequestration: iron reduction using sulfur-bearing waste gas. Chem Geol 217(3):351–364. https://doi.org/10.1016/j.chemgeo.2004.12.018

Sass BM, Farzan H, Prabhakar R, Gerst J, Sminchak J, Bhargava M, Nestleroth B, Figueroa J (2009) Considerations for treating impurities in oxy-combustion flue gas prior to sequestration. Energy Procedia 1(1):535–542. https://doi.org/10.1016/j.egypro.2009.01.071

Seevam PN, Race JM, Downie MJ (2007) Carbon dioxide pipelines for sequestration in the UK: an engineering gap analysis. J Pipeline Eng 6:141–162

Stevens SH, Kuuskraa VA, Gale J, Beecy D (2001) CO_2 injection and sequestration in depleted oil and gas fields and deep coal seams: worldwide potential and costs. Environ Geosci 8(3):200–209. https://doi.org/10.1046/j.1526-0984.2001.008003200.x

Veld K, Mason CF, Leach A (2013) The economics of CO_2 sequestration through enhanced oil recovery. Energy Procedia 37:6909–6919. https://doi.org/10.1016/j.egypro.2013.06.623

Wang C, Li T, Gao H, Zhao J, Li HA (2017) Effect of asphaltene precipitation on CO_2-flooding performance in low-permeability sandstones: a nuclear magnetic resonance study. RSC Adv 7(61):38367–38376. https://doi.org/10.1039/C7RA06392J

Wang J, Ryan D, Anthony EJ, Wigston A (2011) Effects of impurities on geological storage of CO_2. IEAGHG, Stoke Orchard, Cheltenham, UK

Zanganeh P, Dashti H, Ayatollahi S (2015) Visual investigation and modeling of asphaltene precipitation and deposition during CO_2 miscible injection into oil reservoirs. Fuel 160:132–139. https://doi.org/10.1016/j.fuel.2015.07.063

Chapter 5
New Technologies

Abstract Global efforts capturing and storing CO_2 have initiated a variety of trials to utilize CO_2 in enhanced oil recovery (EOR) or enhanced gas recovery (EGR). This Chapter has described the hybrid technologies to introduce the potentials of carbon dioxide capture and storage (CCS) and EOR/EGR. Section 5.1 has reviewed the advanced CO_2 EOR technologies incorporating low salinity water injection (LSWI) or smart water injection (SWI). The CO_2 water-alternating gas (WAG) and carbonated water injection (CWI) are water-based CO_2 EOR, which could be integrated with the LSWI or SWI. Applications of the hybrid CO_2 WAG or CWI combining with LSWI/SWI are reviewed in terms of EOR and CO_2 storage capacity. In Sect. 5.2, the applications of CO_2 injection, which are intended to produce hydrocarbon from shale gas and oil reservoirs and store CO_2 into the depleted shale reservoirs, have been described.

Keywords CO_2 water-alternating gas · Carbonated water injection · Low salinity water injection · Smart water injection · Shale gas and oil

5.1 Salinity Control in CO_2 EOR

A variety of CO_2 injection designs have been developed including continuous CO_2 gas injection, CO_2 WAG, simultaneous water and CO_2 injection, CWI. These designs of CO_2 injection have been investigated to maximize not only EOR effect but CO_2 storage performance (Mahzari et al. 2018; Seyyedi et al. 2018; Shakiba et al. 2016; Foroozesh et al. 2016; Esene et al. 2019; Zhang et al. 2015; Snippe et al. 2020; Kamali and Hussain 2017; Bachu 2016; Hosseininoosheri et al. 2018; Alfi and Hosseini 2016; Peck et al. 2018). Zhang et al. (2015) reviewed the demonstration project of H-59 block pilot on EOR and CO_2 storage in northeast China. In the project, the 20×10^4 tons of CO_2 was injected into the tight oil reservoir from 2008 to 2012. The injected amount of CO_2 was equivalent to the 15.9% of hydrocarbon pore volume (HCPV). The project involved 6 injectors and 25 producers with a spacing of 250–300 m and adopted the miscible CO_2 WAG process. The WAG ratio between 1:2 and 1:5

K. S. Lee et al., *CO_2 Storage Coupled with Enhanced Oil Recovery*,
https://doi.org/10.1007/978-3-030-41901-1_5

was applied to improve sweep efficiency. The project monitored the CO_2 storage and productions. It estimated that nearly 80% of injected CO_2 was stored in reservoir and the rest of injected CO_2 was produced. Along with the CO_2 storage, the oil recovery factor increased from 5 to 14.3% (Fig. 5.1). Alfi and Hosseini (2016) investigated the EOR and CO_2 storage performances in the Cranfield, Mississippi, U.S.A. CO_2 EOR has been deployed in the field since 2008. There were two seismic surveys to monitor the transport of CO_2 in the subsurface. The First survey in 2007 was conducted before CO_2 injection and the other survey in 2010 after considerable CO_2 injection. This study developed history-matched numerical model of CO_2 injection process from 2008 to 2010. From the numerical simulations, it investigated the distribution of injected CO_2 in the formation and, then it predicted the CO_2 EOR and CO_2 storage varying scenarios of injection and production for the next 160 years (Fig. 5.2). It clearly showed that stable structural trapping of CO_2 in the Cranfield and highlighted the post-injection scheme to control the trapping mechanism including structural, geochemical reaction, residual, and mineral trappings.

Recently, the development of novel technology, i.e., Low Salinity Water Injection (LSWI) or Smart Water Injection (SWI), has triggered to develop advanced CCS-CO_2 EOR. Extensive studies have explored the CO_2 EOR, especially CO_2 WAG and CWI, integrating with LSWI or SWI. Because both processes utilize water injection, it is relatively straight-forward to substitute the injection of water stream with LSWI/SWI. Prior to the introduction of the CO_2 WAG and CWI incorporating LSWI/SWI, brief description of LSWI/SWI follows to extend our understanding of hybrid technologies coupled with LSWI/SWI. Next, descriptions of recent researches of hybrid process follow.

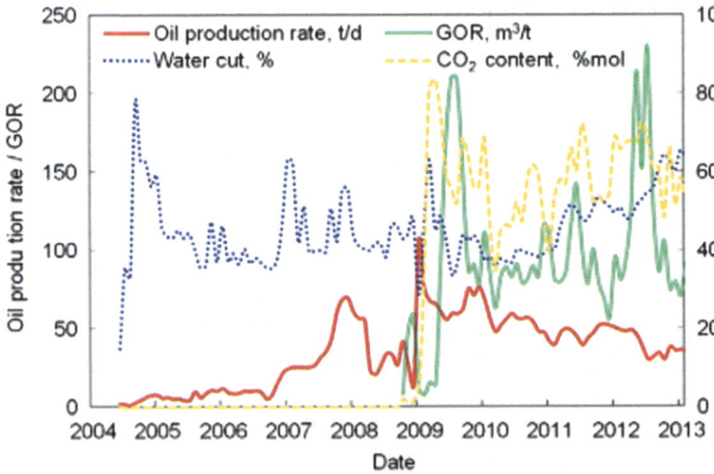

Fig. 5.1 Production data of H-59 pilot production wells (Zhang et al. 2015)

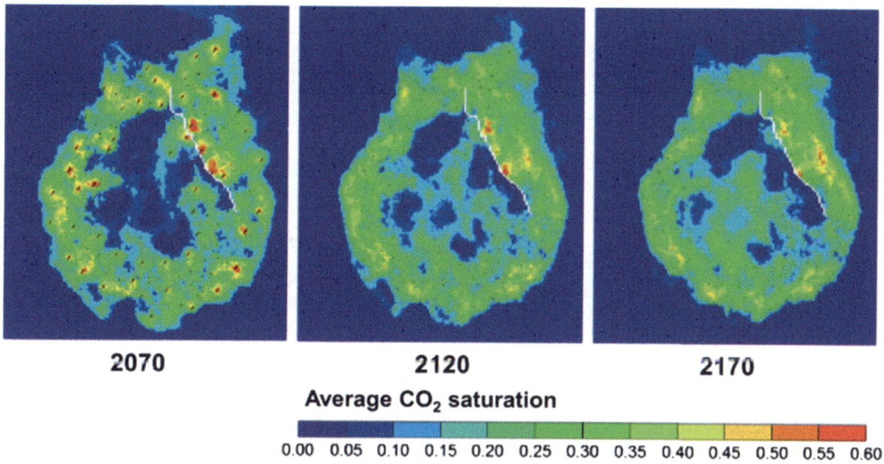

Fig. 5.2 Distribution of injected CO_2 in the formation of Cranfield (Alfi and Hosseini 2016)

5.1.1 LSWI and SWI

The role of ionic composition of brine on the performance of water flood has been explored by many researches (Alotaibi et al. 2011; Austad et al. 2010; Hiorth et al. 2010; Lager et al. 2008; Ligthelm et al. 2009; McGuire et al. 2005; RezaeiDoust et al. 2009; Tang and Morrow 1999; Zhang et al. 2007). It has been reported that lowering salinity or modifying ionic composition of injecting brine would affect the wettability of rocks. As a result, the water injection tuning salinity or ionic composition of brine enhances oil production (Fig. 5.3). The advanced water injection is termed as LSWI/SWI technology. Petroleum industries and researchers designate the technology by a variety of terminologies—LoSal^{TM}, SmartWater, Desinger Waterflood, and Advanced Ion Management (AIM^{TM}). Extensive studies of experiments and field tests have investigated the mechanism behind the LSWI/SWI corresponding to the mineralogy of rocks. Up to date, many theories—fine migration, in situ generation of surfactant, multi-component ionic exchange (MIE), salting-in effect, potential determining ions (PDI) (Fig. 5.4), electrical double layer (EDL) expansion, pH increase, mineral dissolution, surface charge, adhesion energy of crude oil/brine/rock—have been proposed to explain the observations of experiments and pilot tests (Alotaibi et al. 2011; Austad et al. 2010; Hiorth et al. 2010; Lager et al. 2008; Ligthelm et al. 2009; McGuire et al. 2005; RezaeiDoust et al. 2009; Tang and Morrow 1999; Zhang et al. 2007). Though the unanimous mechanism of LSWI/SWI has not revealed yet, the observations of experiments and field tests clearly demonstrate that injecting brine, which has low salinity or potential determining ions (Ca^{2+}, Mg^{2+}, and SO_4^{2-}), modifies the wettability of sandstone or carbonate rocks due to geochemical reactions.

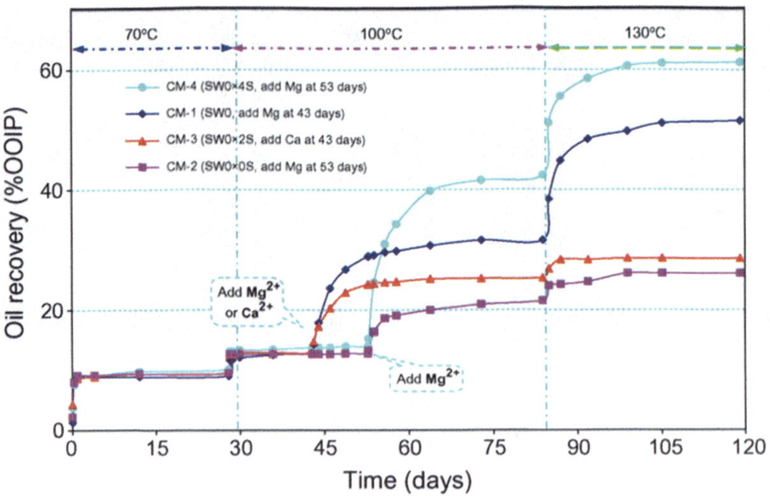

Fig. 5.3 Spontaneous imbibition test of seawater injection modifying ionic composition (Zhang et al. 2007)

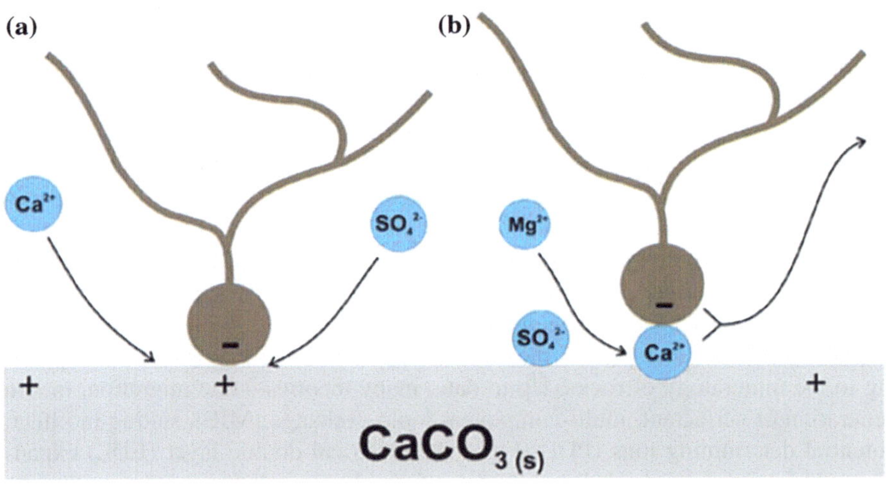

Fig. 5.4 Description of potential determining ions mechanisms in carbonate reservoir (Zhang et al. 2007)

5.1.2 Advanced CO_2 WAG Integrated with LSWI/SWI

Extensive studies of CO_2 WAG have investigated to adjust the chemical condition of injected water slug inspired by LSWI/SWI technology (Dang et al. 2014; Wilson 2015; Kumar and Mandal 2017; Afzali et al. 2018; Al-Saedi et al. 2019a, b; Kulkarni and Rao 2005; Teklu et al. 2014, 2016; Al-Saedi and Flori 2019; Al-Shalabi et al.

2016; Naderi and Simjoo 2019). Because the hybrid low salinity-CO_2 WAG (LS-CO_2 WAG) process is a combined process between CO_2 WAG and LSWI, it could overcome the current challenges associated with the application of LSWI or CO_2 WAG for green and brown oil fields (Dang et al. 2014; Wilson 2015). In addition, the hybrid process introduces a number of advantages including mobility control, oil swelling, density change, miscibility, viscosity reduction, water blockage for high permeable zone, and wettability modification effect. Dang et al. (2014) numerically modeled the hybrid LS-CO_2 WAG process incorporating the mechanisms of CO_2 WAG and LSWI processes. Developing the model considering multi-phase and multi-component transport, equation of state (EOS), and geochemical reactions, they simulated and compared the various processes of conventional water flood, LSWI, CO_2 gas injection, CO_2 WAG, LS-CO_2 WAG for a core-scaled system. The hybrid process clearly produced ultimate oil recovery factor and higher oil production compared to the conventional water flood and LSWI due to miscibility effect. In comparison to the CO_2 gas injection and CO_2 WAG, LS-CO_2 WAG produced oil much faster in the early stage of WAG cycles. The core-scaled simulations identified the effects of solubility of CO_2 in brine, dissolution of carbonate minerals, ion exchange, and wettability alteration on the performance of LS-CO_2 WAG process. Then, this study extended the simulation of LS-CO_2 WAG to the field scale model of Brugge field. This simulation study summarized that the success of LS-CO_2 WAG process depended on the heterogeneity, initial wettability, clay, mineralogy, brine composition, reservoir pressure and temperature for CO_2 miscibility, and WAG parameters. Because these factors are key parameters for at least LSWI or CO_2 WAG processes, successful LSWI-CO_2 WAG project should satisfy screening criteria of both LSWI and CO_2 WAG processes.

A couple of studies experimentally investigated the LS-CO_2 WAG process with coreflooding, contact angle, and IFT measurement tests (Teklu et al. 2014, 2016). The contact angle measurement test indicated that low salinity water condition decreased contact angle improving wettability. In addition to wettability modification due to low salinity water condition, the effect of salinity-dependent CO_2 solubility in brine on the additional wettability modification was investigated. Generally, CO_2 solubility in brine is a function of pressure, temperature, and salinity. The solubility of CO_2 increases with a decrease in salinity, which is called salting-out phenomenon (Fig. 5.5). The study explained that CO_2 solubility increment with low-salinity water condition reduced IFT between CO_2 and brine (Fig. 5.6). As a result, the IFT between oil and brine decreased when the brine was saturated with CO_2. The reduction in IFT between oil and CO_2-saturated brine caused additional wettability modification effect during the LS-CO_2 WAG process. In the coreflooding test, the study injected the LS-CO_2 WAG following seawater injection. The test clearly showed an increasing oil recovery of LS-CO_2 WAG comparing to the seawater injection.

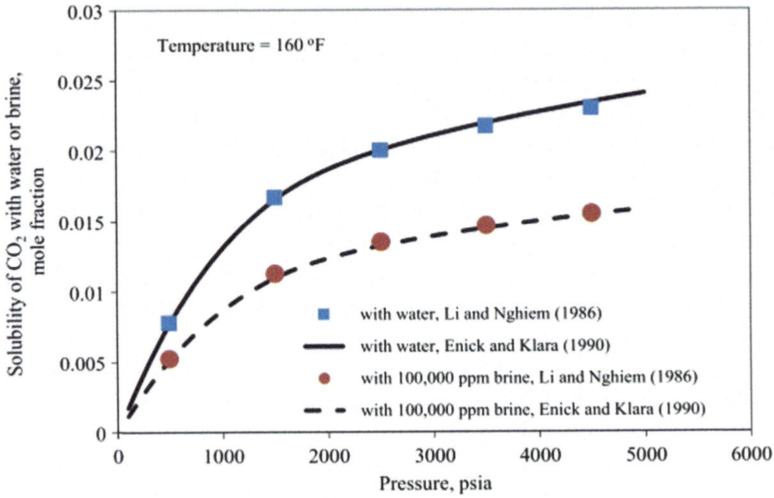

Fig. 5.5 CO_2 solubility in fresh and 100,000 ppm salinity brine (Teklu et al. 2016)

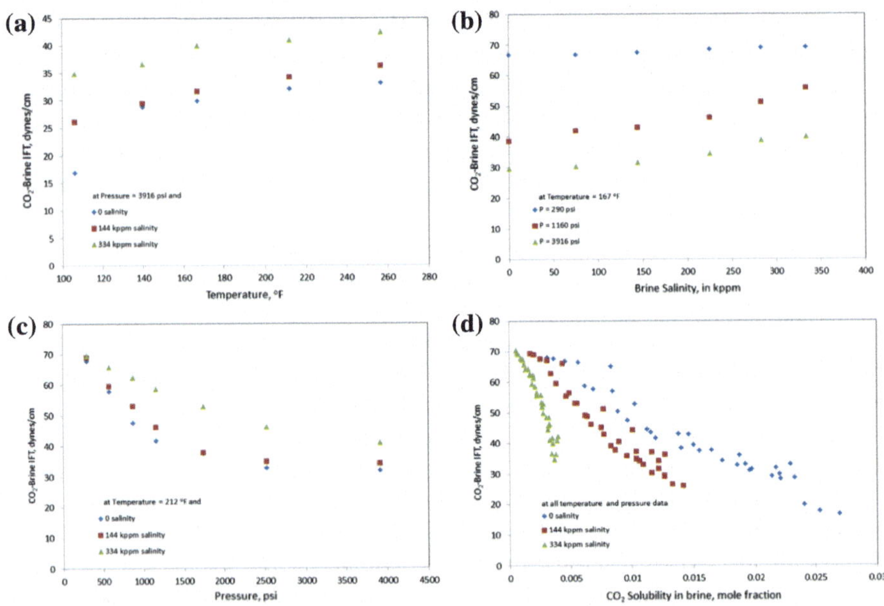

Fig. 5.6 IFT of brine and CO_2 as a function of **a** temperature, **b** salinity, **c** pressure, **d** CO_2 solubility in brine (Teklu et al. 2016)

Al-Shalabi et al. (2016) numerically investigated the various scenarios of LS-immiscible/miscible CO_2 WAG process including constant WAG, tapered WAG, and simultaneous WAG (SWAG). During the test, the salinity of injecting water slug has been varied during the LS-WAG process. The results of simulation showed SWAG outperformed all other combinations of WAG processes in terms of oil recovery factor and utilization factor of CO_2, which is defined as the thousand standard cubic feet of gas required to produce one barrel of oil. The study addressed the synergetic effects of LS-CO_2 WAG when the process is under miscible condition. When miscible CO_2 injection displaced the residual oil saturation, the low salinity water effectively boosted the oil production rate due to improving wettability modification toward water-wet condition. They also conducted a fractional flow curve analysis corresponding to the water/oil and water/solvent curves. The analysis evaluated the economic amount of solvent required to achieve ultimate oil recovery factor (Fig. 5.7). The total mobility curves were used to calculate the effective displacement mobility ratio, which enables to estimate the optimum solvent usage. It was drawn that higher and faster oil recovery with less amount of solvent usage could be achieved for LS-CO_2 WAG process. Kumar and Mandal (2017) also discussed the technical issues and challenges for the deployment of WAG process utilizing chemical agents and low salinity water.

A series of studies (Al-Saedi et al. 2019a, b; Al-Saedi and Flori 2019) have investigated the LS-CO_2 WAG process through coreflooding, contact angle, spontaneous imbibition tests. Al-Saedi et al. (2019a) observed that higher tertiary oil recovery of coreflooding was observed in LS-CO_2 WAG over CO_2 WAG by approximately 4%. Higher pressure drop was also monitored due to the fines mobilization during LS-CO_2 WAG process. These observations were already confirmed by LSWI studies (Tang and Morrow 1999; RezaeiDoust et al. 2009; Austad et al. 2010). After testing a number of brine compositions to select optimum low salinity brine, Al-Saedi and Flori (2019) used the optimum brine in the water injection stream of LS-CO_2 WAG. In the coreflooding test, the LS-CO_2 WAG was operated with three cycles of WAG following brine injection. It was clearly observed that tertiary recovery of LS-WAG produced 24.5% higher oil recovery than secondary recovery of brine injection. Al-Saedi et al. (2019b) also observed the increasing oil recovery of LS-CO_2 WAG compared to the conventional CO_2 WAG (Fig. 5.8). The study explained that modifying ionic composition of brine in the CO_2 WAG process contributed wettability modification. The more CO_2 dissolution in oil, the improved sweep efficiency is observed to result in higher oil recovery. The study noted that this promising EOR effect could be achievable in the laboratory condition. However, the existence of heterogeneity, fractures, and other reservoir characteristics could limit the significant EOR performance behind LS-CO_2 WAG in the real field.

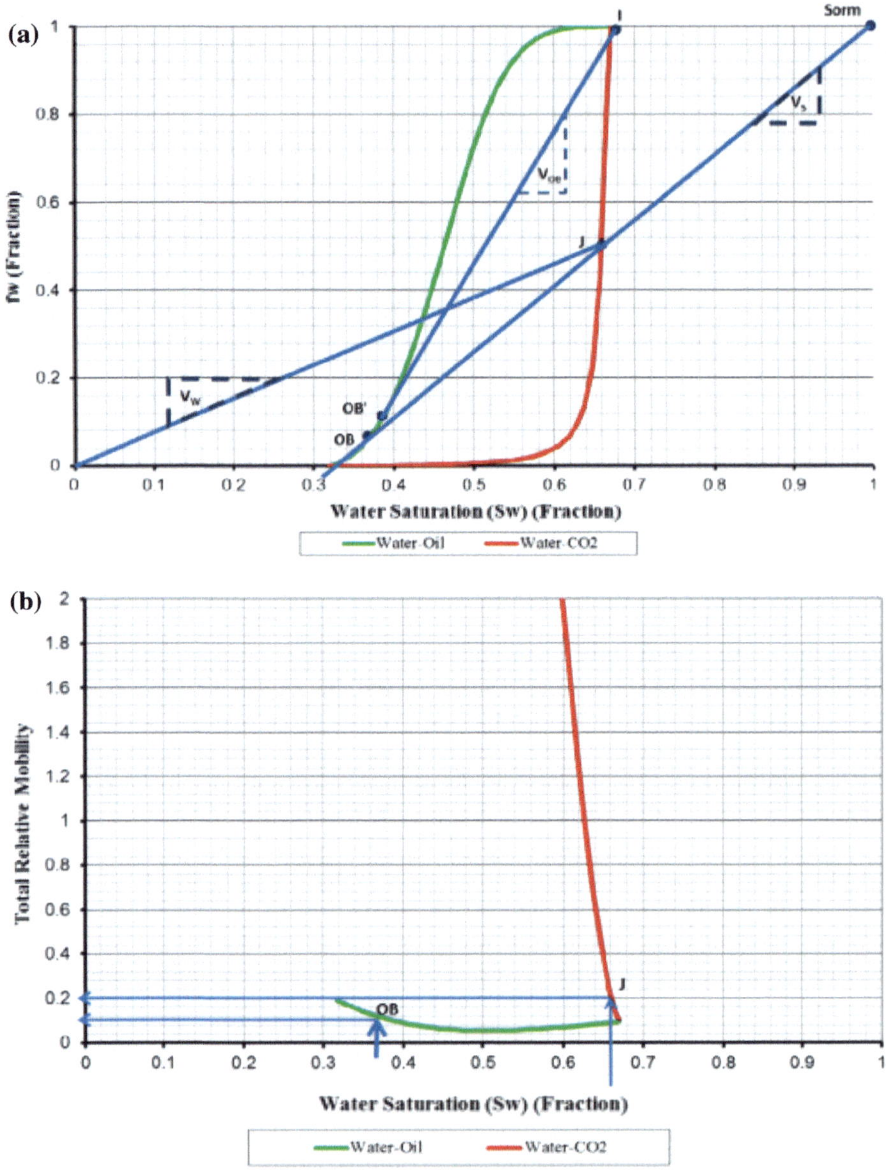

Fig. 5.7 Analysis of LSWI-CO_2 WAG—**a** fractional flow curves and **b** total mobility curves (Al-Shalabi et al. 2016)

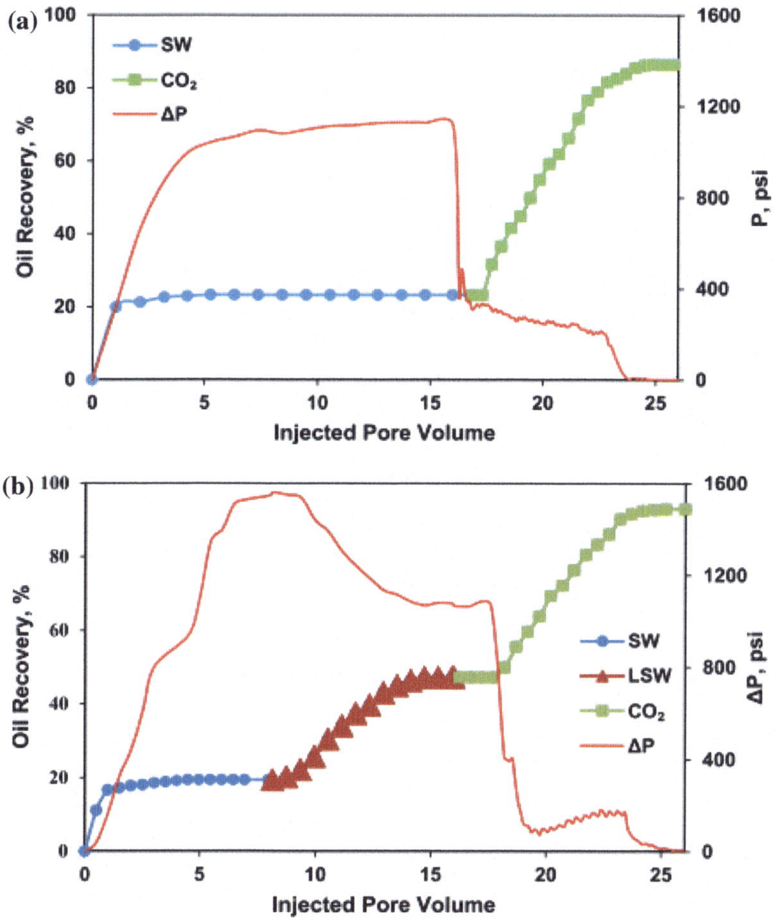

Fig. 5.8 Oil recoveries of coreflooding—**a** CO_2 WAG and **b** LS-CO_2 WAG (Al-Saedi et al. 2019b)

5.1.3 Advanced CWI Integrated with LSWI/SWI

The water injection of WAG process mitigates the CO_2 gas breakthrough and improves sweep efficiency of the process. However, a significant water-blocking effect occurs during CO_2 WAG and it hinders contacting between a large amount of oil and injected CO_2 (Shu et al. 2014). In the water-wet reservoirs, the water-blocking effect reduces the microscopic displacement efficiency (Patel et al. 1987; Land 1990; Lin and Huang 1990; Shu et al. 2014). The CWI is an alternative CO_2 EOR, which is a water-based CO_2 injection. Shu et al. (2014) carried out an experimental research with coreflooding tests examining the potential of pre-flush CWI followed by CO_2 gas injection (Fig. 5.9). The study concluded that pre-flush CWI eliminated water-blocking effect by the mass transfer of CO_2 from the carbonate water to oil. Another

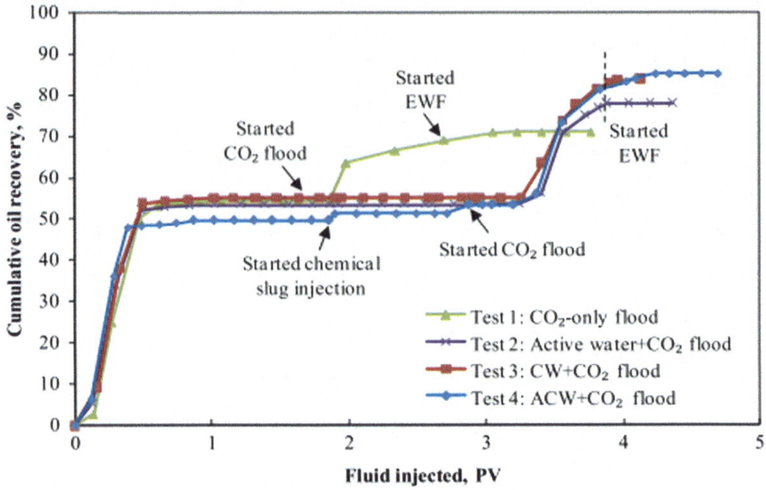

Fig. 5.9 Oil recoveries of CO_2 gas injection with and without pre-flush CWI (Shu et al. 2014)

conclusion of pre-flush CWI application was to promote the contacting between following CO_2 gas stream and trapped oil. In addition to these effects, CWI enhances oil production attributed to the oil swelling, oil viscosity reduction, and oil density reduction effects. These changes in oil properties were caused by CO_2 mass transfer between carbonated water and oil (Foroozesh et al. 2016; Shakiba et al. 2016; Mahzari et al. 2018; Seyyedi et al. 2018). Along with the CO_2 mass transfer, CWI resulted in the composition variations of in situ oil extracting intermediate oil components into the gaseous phase. It made in situ oil to be heavier than original oil (Fig. 5.10).

In addition to the EOR effect, CWI also stably sequestrates CO_2 in the reservoir formations (Foroozesh et al. 2016; Shakiba et al. 2016; Mahzari et al. 2018; Esene

Fig. 5.10 Composition variations of in situ oil after CWI process (Mahzari et al. 2018)

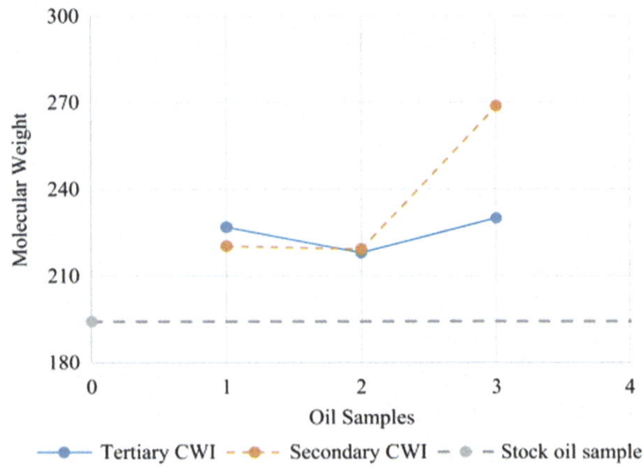

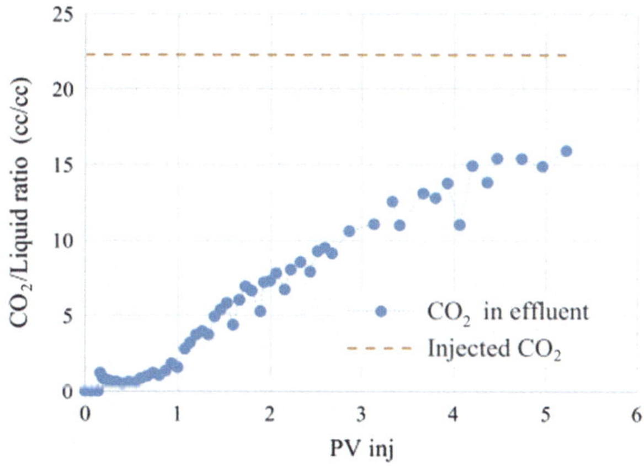

Fig. 5.11 Concentrations of CO_2 in injected and produced liquids during CWI (Mahzari et al. 2018)

et al. 2019). Many researches of CWI were mainly interested in the solubility trapping mechanism rather than other structural trapping and residual trapping mechanisms. Because CWI injects CO_2-saturated brine with high pressure condition, the injected CO_2 is prone to be stored by solubility trapping. Mahzari et al. (2018) experimentally analyzed the concentration of CO_2 in the produced liquid to monitor the CO_2 retention by trapping in the porous medium (Fig. 5.11). The difference between injected and produced CO_2 indicated the CO_2 sequestration capacity of the CWI process. Esene et al. (2019) estimated the CO_2 storage of CWI in the sub-model of North Sea field. Avoiding fracturing and leakage risks, the implementation of CWI stored significant amount of injected CO_2 up to 82.56%. Because the CWI requires less CO_2 to be injected compared to the CO_2 gas injection, the injected CO_2 can be efficiently stored via solubility trapping mechanism.

Recently, a number of studies have investigated the hybrid process between CWI and LSWI/SWI, hereafter referred as LS-CWI or Smart-CWI. The studies (Kilybay et al. 2016, 2017) conducted NMR test, IFT measurement, Zeta potential measurement, micro-emulsion test, ionic composition analysis, and coreflooding test. The Zeta potential test and IFT measurement were carried out to prepare and formulate the optimum injecting brine of Smart-CWI. Coreflooding tests evaluated the Smart-CWI injection following SWI injection. Smart-CWI recovered additional oil from 5.7 to 13.6% over SWI. Both NMR and ionic composition analysis tests observed the changes in the pore size distribution and the ionic composition of produced brine. The studies explained that the changes were caused by the rock dissolution. The conclusion of the studies was made that EOR effect behind Smart-CWI was attributed to the geochemical reactions of rock dissolution and ion-exchange, oil viscosity reduction, oil swelling, and reconnection of oil ganglia.

Hamouda and Bagalkot (2019) also carried out CWI study experimentally and numerically to quantify the relationship between IFT of carbonated brine/n-Decane and salt. It also measured the CO_2 diffusion coefficient describing the inter-phase transfer of CO_2 by varying salt concentrations. During the tests, the concentrations of Mg^{2+} and SO_4^{2-} in the carbonated brine varied. Reduction in IFT was observed when the carbonated brine had Mg^{2+} (Fig. 5.12a). It was explained that high hydration energy of Mg^{2+} reduced more IFT when the cation adsorbed on the interface. In the case of SO_4^{2-} exhibiting low hydration energy, more interphase CO_2 transfer into the n-Decane occurred and more IFT was reduced (Fig. 5.12b). Nowrouzi et al. (2019) also measured IFT of brine and crude oil at a temperature of 75° and a pressure 2,000 psi by varying the concentrations of salts and dissolved CO_2 in brine. It was observed that IFT decreased as the brine contained more ions and soluble CO_2 (Fig. 5.13). The reason behind the IFT reduction due to salts was explained with the kinetic adsorption of boundary active molecules on the surface of the liquid. Another observation reducing IFT due to the dissolved CO_2 in brine was attributed to the in situ surfactant generation and oil swelling by the interphase transfer of CO_2. The study also showed the salt-type-dependent reduction of IFT between carbonated water and crude oil. While previous studies extensively quantified the mechanism

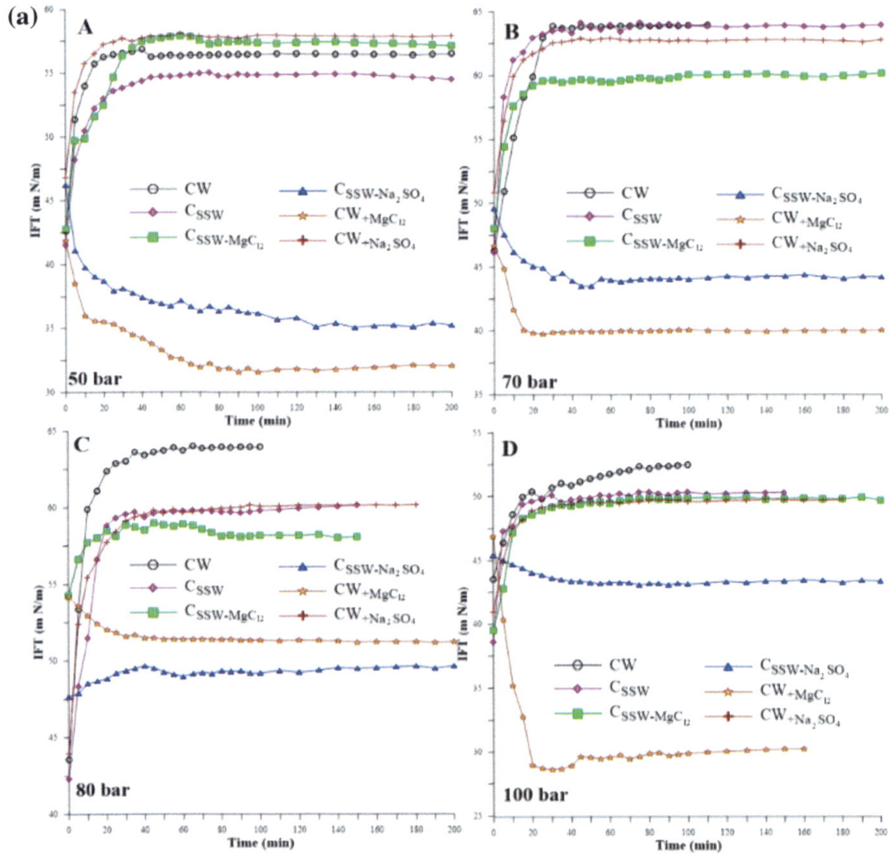

Fig. 5.12 Results of changes in **a** IFT and **b** diffusion coefficient of CO_2 by varying salt (Mg^{2+} and SO_4^{2-}) (Hamouda and Bagalkot 2019)

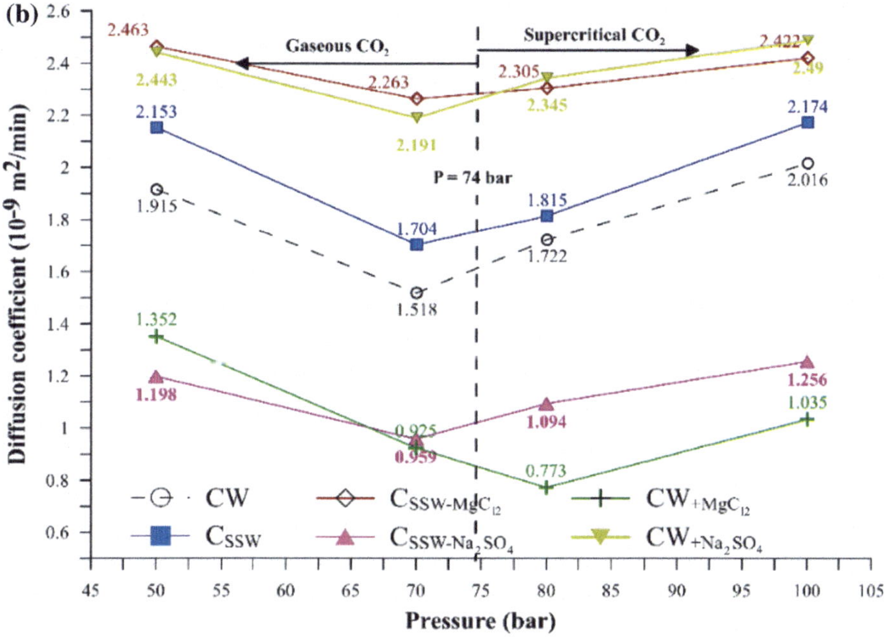

Fig. 5.12 (continued)

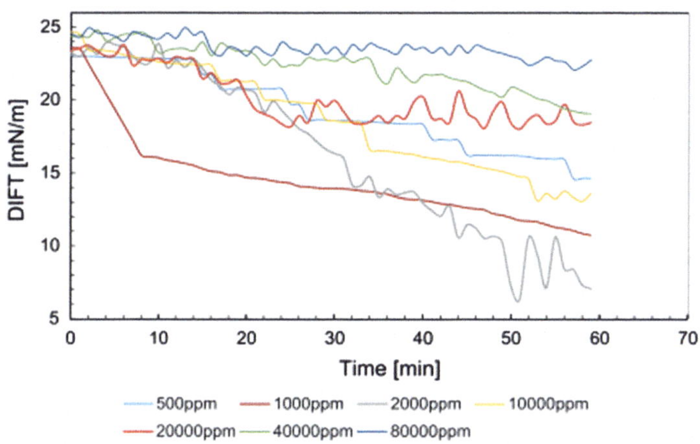

Fig. 5.13 Variations of IFT of carbonated water and crude oil at different concentrations of $MgSO_4$ (Nowrouzi et al. 2019)

of hybrid LS-CWI or Smart-CWI, field- or pilot-scaled investigations are required for the successful deployment in the field. It is obvious that the hybrid process is a promising technology to enhance both oil production and CO_2 storage.

5.2 Unconventional Reservoirs

Shale reservoirs have been known as an enormous hydrocarbon source. Commercial success of hydrocarbon production from shale formations is mainly due to the development in multi-fracturing and horizontal drilling technologies (Rani et al. 2019; Zhou et al. 2019). The trials of CO_2 injection into the low permeable shale formation have been considered to enhance hydrocarbon production. For the shale gas reservoir, 40–50% of initial gas, mainly CH_4, is stored by the adsorption mechanism and the gas stored by the adsorption could be recovered by CO_2 injection. Because CO_2 shows 1.7–8.1 times higher adsorption potential over CH_4, injected CO_2 attaches to the rock surface detaching CH_4 (Zhou et al. 2019). As a result, CO_2 injection enhances shale gas production. Another benefit of CO_2 injection is to sequestrate CO_2 in shale formation. Because the tight shale formations typically function as both reservoir rock and cap-rock, they are attractive targets storing CO_2 after CO_2 injection. In addition to the general CO_2 storage mechanisms, additional storing mechanism via CO_2 adsorption becomes of importance in the low permeable shale formation. Because CO_2 shows significant adsorption capacity in the shale formation, the adsorption plays a critical role in the CO_2 gas storage. The adsorptions of CO_2 as well as CH_4 occur on the organic matter in the micro- and nano-pores (Fig. 5.14). A small amount of clay deposited is also responsible for the gas adsorption (Fig. 5.14). Therefore, the adsorptive capacity of shale is sensitive to the composition of the shale formation

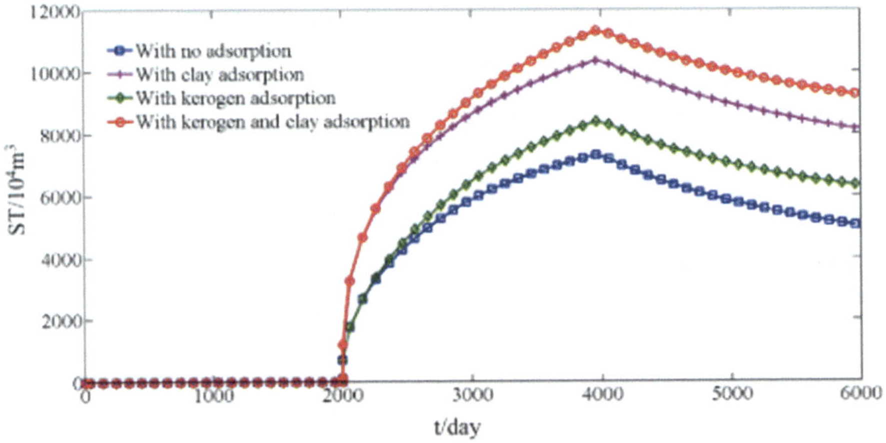

Fig. 5.14 The effect of adsorption on CO_2 storage capacity (Wu et al. 2019)

Fig. 5.15 Isothermal adsorption behavior of CO_2 in the shale samples before and after supercritical CO_2 exposure (Zhou et al. 2018)

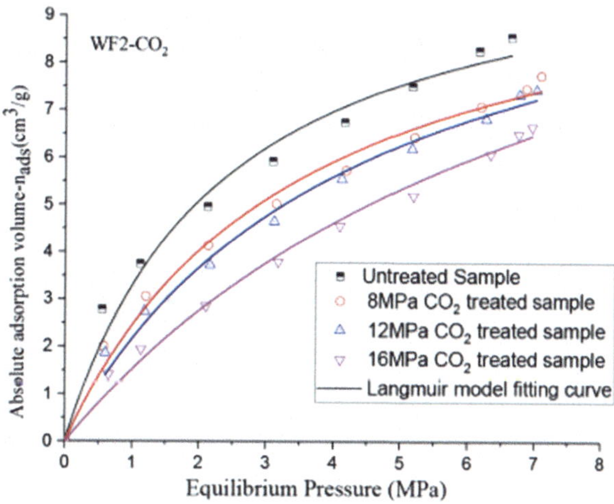

as well as pore structure (Rani et al. 2019). Following studies have described the researches of CO_2 injection to enhance gas production and sequestrate CO_2 in the shale formations (Zhou et al. 2018; Goodman et al. 2019; Li and Elsworth 2019; Merey 2019; Myshakin et al. 2019; Rani et al. 2019; Wu et al. 2019; Zhou et al. 2019).

Majority of the studies have focused on CH_4 and CO_2 adsorption behaviors to evaluate the EOR and CO_2 storage capacity. Zhou et al. (2018) investigated the effect of supercritical CO_2 on the adsorption behaviors. Because of the high pressure and temperature conditions of the reservoir, the injected CO_2 exhibits the supercritical characteristics. Isotherm adsorption experiments of CO_2 and CH_4 were conducted using the shale rock samples from the Longmaxi and Wufeng formations, Sichuan Basin, China. The experiments compared the adsorption behaviors when the rock samples were exposed to the supercritical CO_2 or not. The experimental observations showed a decrease in the adsorption when the rock samples were treated with supercritical CO_2 (Fig. 5.15). It was explained that supercritical CO_2 affected the pore structures, mineral composition, and surface chemistry of shale rocks. These observations imply less sequestration of CO_2 via adsorption mechanism in shale formations.

Li and Elsworth (2019) numerically compared the injections of pure CO_2, pure N_2, and mixture $N_2 + CO_2$ in terms of shale gas recovery and CO_2 sequestration. Because the injected N_2 efficiently lowered the partial pressure of CH_4 in the porous media, N_2 injection resulted in higher shale gas production (Fig. 5.16a). The injection of CO_2 gas recovered the CH_4 via competitive adsorption (Fig. 5.16a). While injections of both CO_2 and N_2 recovered shale gas, N_2 injection was more effective in enhanced shale gas recovery. In terms of CO_2 sequestration, injection of pure CO_2 gas stream showed less performance than the injection of mixture $CO_2 + N_2$ (Fig. 5.16b). These observations indicated that co-optimization of EOR and CO_2 sequestration required the optimum recipes of mixture $N_2 + CO_2$.

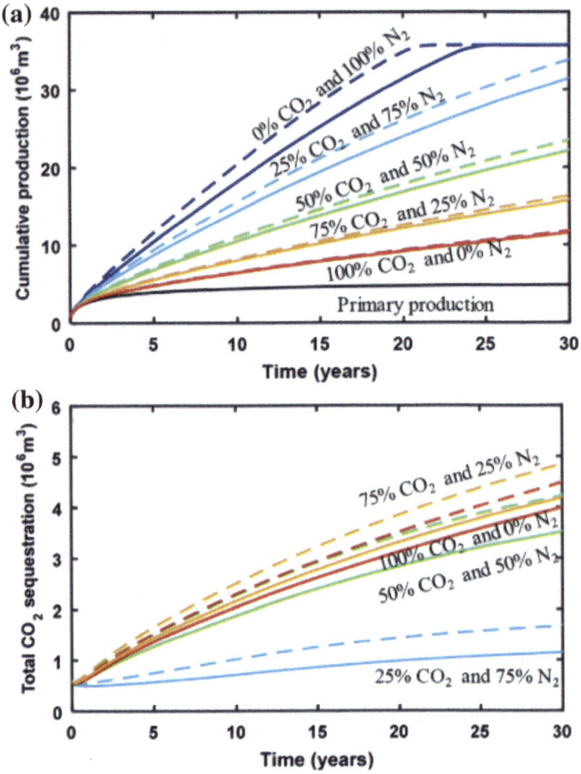

Fig. 5.16 Results of numerical simulations of pure CO_2, pure N_2, and mixture $CO_2 + N_2$ injections: **a** shale gas production and **b** CO_2 sequestration (Li and Elsworth 2019)

While the role of adsorption behavior on the fluid transport in the shale gas formation has been explored by many researches, investigating an interaction in the CO_2/rock/connate water system is limited. Goodman et al. (2019) experimentally evaluated the effect of CO_2 dissolution into connate water on the physical and chemical alterations of the carbonate-rich Utica Shale during the CO_2 injection. Experiments observed the carbonate dissolution and precipitation reactions due to CO_2 dissolution in water through a series of the tests including attenuated total Reflectance-Fourier transform infrared spectroscopy (ATR-FT-IR), scanning electron microscopy (SEM), and fluid chemical analysis. The tests observed the changes in carbonate chemistry indicating modifications in pore size distribution. These observations implied that CO_2 injection causing the chemical reactions would modify flow pathways in the carbonate-rich shale system and might affect CO_2 storage and hydrocarbon recovery.

Recently, the application of CO_2 injection has been investigated in the shale oil reservoirs (Pranesh 2018; Jia et al. 2019; Lashgari et al. 2019; Lee et al. 2019; Fakher and Imqam 2020). In the view of shale oil productions, a number of factors are important such as interactions between crude oil and CO_2, molecular diffusion, heterogeneity, and nano-confinement effect (Jia et al. 2019). CO_2 injection into the shale oil reservoirs also sequestrates CO_2 in the subsurface formations (Pranesh 2018; Jia et al. 2019; Lashgari et al. 2019; Lee et al. 2019; Fakher and Imqam 2020). The injected CO_2 could be sequestrated in shale oil formation by adsorption, hysteresis, retention, structural trapping, chemical reactions, migration of CO_2 into the aquifer below, and leakage of CO_2 to thief zone (Pranesh 2018).

Fakher and Imqam (2020) conducted coreflooding of cyclic CO_2 injection and measured the enhanced shale oil recovery and CO_2 storage via adsorption. The experiment controlled the factors of experimental systems such as injection pressure, soaking time, injection cycles, CO_2 injection phase, temperature, and fractures. Conditions of the system including high injection pressure, more soaking time, more injection cycles, fractures, higher temperature, and supercritical phase of injected CO_2 enhanced oil production from shale formations (Fig. 5.17). CO_2 sequestration

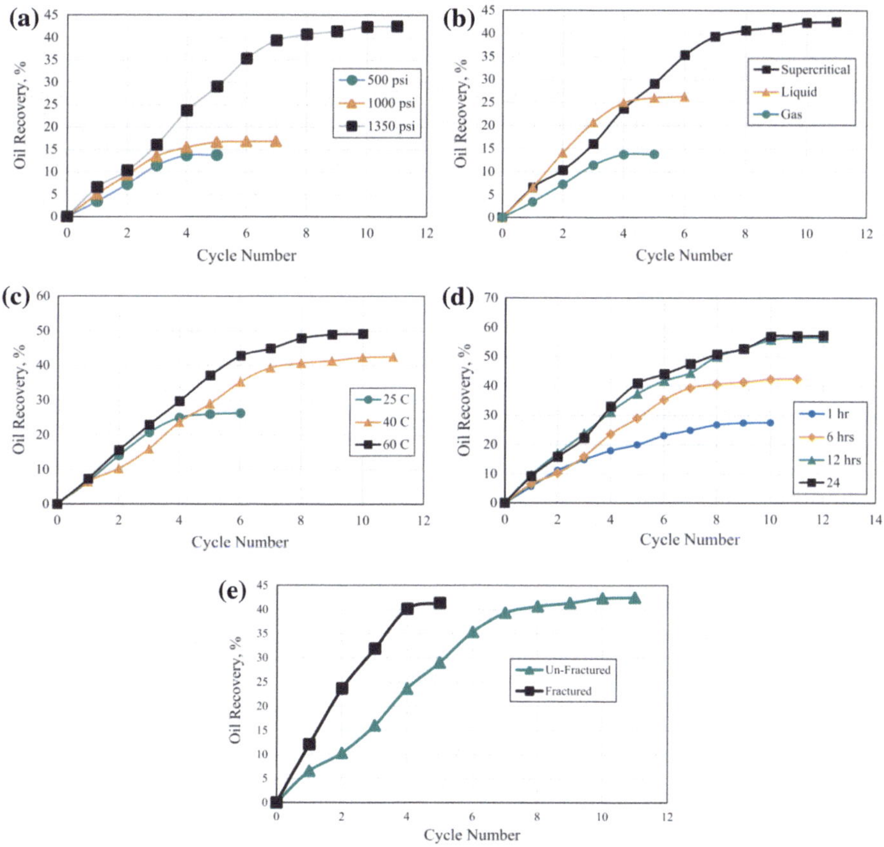

Fig. 5.17 Shale oil recovery of coreflooding by varying **a** injection pressure, **b** CO_2 injection phase, **c** temperature, **d** soaking time, and **e** stimulation by fracture for cyclic CO_2 injection (Fakher and Imqam 2020)

by adsorption behavior was also monitored and conditions of high injection pressure, small shale particle size, and low temperature condition were determined as favorable conditions to sequestrate CO_2 in shale oil formation. Up to date, extensive investigations of CO_2 injection mainly focused on the shale gas reservoirs rather than shale oil reservoirs. Previous studies sufficiently demonstrated EOR potential and CO_2 storage capacity of shale oil formations. Therefore, CO_2 injection in shale oil formations would be the next CCS-CO_2 technology in future.

References

Afzali S, Rezaei N, Zendehboudi S (2018) A comprehensive review on enhanced oil recovery by water alternating gas (WAG) injection. Fuel 227:218–246. https://doi.org/10.1016/j.fuel.2018.04.015

Al-Saedi HN, Flori RE (2019) Novel coupling smart water-CO_2 flooding for sandstone reservoirs. Petrophysics 60(04):525–535

Al-Saedi HN, Al-Jaberi SK, Flori RE, Al-Bazzaz W, Long Y (2019a) A new design of low salinity-CO_2-different chemical matters. In: Paper presented at the Abu Dhabi international petroleum exhibition and conference, Abu Dhabi, UAE, 11 Nov 2019. https://doi.org/10.2118/197118-ms

Al-Saedi HN, Long Y, Flori RE, Bai B (2019b) Coupling smart seawater flooding and CO_2 flooding for sandstone reservoirs: smart seawater alternating CO_2 flooding (SMSW-AGF). Energy Fuels 33(10):9644–9653. https://doi.org/10.1021/acs.energyfuels.9b02289

Al-Shalabi EW, Sepehrnoori K, Pope G (2016) Numerical modeling of combined low salinity water and carbon dioxide in carbonate cores. J Petrol Sci Eng 137:157–171. https://doi.org/10.1016/j.petrol.2015.11.021

Alfi M, Hosseini SA (2016) Integration of reservoir simulation, history matching, and 4D seismic for CO_2-EOR and storage at Cranfield, Mississippi, USA. Fuel 175:116–128. https://doi.org/10.1016/j.fuel.2016.02.032

Alotaibi MB, Nasr-El-Din HA, Fletcher JJ (2011) Electrokinetics of limestone and dolomite rock particles. SPE Reserv Eval Eng 14(5):594–603. https://doi.org/10.2118/148701-PA

Austad T, Rezaeidoust A, Puntervold T (2010) Chemical mechanism of low salinity water flooding in sandstone reservoirs. In: Paper presented at the SPE improved oil recovery symposium, Tulsa, Oklahoma, USA, 24–28 April. https://doi.org/10.2118/129767-ms

Bachu S (2016) Identification of oil reservoirs suitable for CO_2-EOR and CO_2 storage (CCUS) using reserves databases, with application to Alberta, Canada. Int J Greenhouse Gas Control 44:152–165. https://doi.org/10.1016/j.ijggc.2015.11.013

Dang CTQ, Nghiem LX, Chen Z, Nguyen NTB, Nguyen QP (2014) CO_2 low salinity water alternating gas: a new promising approach for enhanced oil recovery. In: Paper presented at the SPE improved oil recovery symposium, Tulsa, Oklahoma, USA, 12 April 2014. https://doi.org/10.2118/169071-ms

Esene C, Zendehboudi S, Aborig A, Shiri H (2019) A modeling strategy to investigate carbonated water injection for EOR and CO_2 sequestration. Fuel 252:710–721. https://doi.org/10.1016/j.fuel.2019.04.058

Fakher S, Imqam A (2020) Application of carbon dioxide injection in shale oil reservoirs for increasing oil recovery and carbon dioxide storage. Fuel 265:116944. https://doi.org/10.1016/j.fuel.2019.116944

Foroozesh J, Jamiolahmady M, Sohrabi M (2016) Mathematical modeling of carbonated water injection for EOR and CO_2 storage with a focus on mass transfer kinetics. Fuel 174:325–332. https://doi.org/10.1016/j.fuel.2016.02.009

Goodman A, Sanguinito S, Tkach M, Natesakhawat S, Kutchko B, Fazio J, Cvetic P (2019) Investigating the role of water on CO_2-Utica Shale interactions for carbon storage and shale gas extraction activities—evidence for pore scale alterations. Fuel 242:744–755. https://doi.org/10.1016/j.fuel.2019.01.091

Hamouda AA, Bagalkot N (2019) Effect of salts on interfacial tension and CO_2 mass transfer in carbonated water injection. Energies 12(4). https://doi.org/10.3390/en12040748

Hiorth A, Cathles LM, Madland MV (2010) The impact of pore water chemistry on carbonate surface charge and oil wettability. Transp Porous Media 85(1):1–21. https://doi.org/10.1007/s11242-010-9543-6

Hosseininoosheri P, Hosseini SA, Nuñez-López V, Lake LW (2018) Impact of field development strategies on CO_2 trapping mechanisms in a CO_2–EOR field: a case study in the permian basin (SACROC unit). Int J Greenhouse Gas Control 72:92–104. https://doi.org/10.1016/j.ijggc.2018.03.002

Jia B, Tsau J-S, Barati R (2019) A review of the current progress of CO_2 injection EOR and carbon storage in shale oil reservoirs. Fuel 236:404–427. https://doi.org/10.1016/j.fuel.2018.08.103

Kamali F, Hussain F (2017) Field-scale simulation of CO_2 enhanced oil recovery and storage through SWAG injection using laboratory estimated relative permeabilities. J Petrol Sci Eng 156:396–407. https://doi.org/10.1016/j.petrol.2017.06.019

Kilybay A, Ghosh B, Thomas NC, Aras P (2016) Hybrid EOR technology: carbonated water and smart water improved recovery in oil wet carbonate formation. In: Paper presented at the SPE annual caspian technical conference and exhibition, Astana, Kazakhstan, 1 Nov 2016. https://doi.org/10.2118/182567-ms

Kilybay A, Ghosh B, Thomas NC, Sulemana NT (2017) Hybrid EOR technology: carbonated water-smart water flood improved recovery in oil wet carbonate formation: Part-II. In: Paper presented at the SPE oil and gas India conference and exhibition, Mumbai, India, 4 April 2017. https://doi.org/10.2118/185321-ms

Kulkarni MM, Rao DN (2005) Experimental investigation of miscible and immiscible water-alternating-gas (WAG) process performance. J Petrol Sci Eng 48(1):1–20. https://doi.org/10.1016/j.petrol.2005.05.001

Kumar S, Mandal A (2017) A comprehensive review on chemically enhanced water alternating gas/CO_2 (CEWAG) injection for enhanced oil recovery. J Petrol Sci Eng 157:696–715. https://doi.org/10.1016/j.petrol.2017.07.066

Lager A, Webb KJ, Black CJJ, Singleton M, Sorbie KS (2008) Low salinity oil recovery—an experimental investigational. Petrophysics 49(1):28–35

Land CS (1990) Calculation of the water-blocking effect in tertiary miscible displacement. In: Society of petroleum engineers, p 30

Lashgari HR, Sun A, Zhang T, Pope GA, Lake LW (2019) Evaluation of carbon dioxide storage and miscible gas EOR in shale oil reservoirs. Fuel 241:1223–1235. https://doi.org/10.1016/j.fuel.2018.11.076

Lee JH, Jeong MS, Lee KS (2019) Incorporation of multi-phase solubility and molecular diffusion in a geochemical evaluation of the CO_2 huff-n-puff process in liquid-rich shale reservoirs. Fuel 247:77–86. https://doi.org/10.1016/j.fuel.2019.03.033

Li Z, Elsworth D (2019) Controls of CO_2–N_2 gas flood ratios on enhanced shale gas recovery and ultimate CO_2 sequestration. J Petrol Sci Eng 179:1037–1045. https://doi.org/10.1016/j.petrol.2019.04.098

Ligthelm DJ, Gronsveld J, Hofman J, Brussee N, Marcelis F, van der Linde H (2009) Novel waterflooding strategy by manipulation of injection brine composition. In: Paper presented at the EUROPEC/EAGE conference and exhibition, Amsterdam, The Netherlands, 8–11 June. https://doi.org/10.2118/119835-ms

Lin EC, Huang ETS (1990) The effect of rock wettability on water blocking during miscible displacement. SPE Reser Eng 5(02):205–212. https://doi.org/10.2118/17375-PA

Mahzari P, Tsolis P, Sohrabi M, Enezi S, Yousef AA, Eidan AA (2018) Carbonated water injection under reservoir conditions; in-situ WAG-type EOR. Fuel 217:285–296. https://doi.org/10.1016/j.fuel.2017.12.096

McGuire PL, Chatham JR, Paskvan FK, Sommer DM, Carini FH (2005) Low salinity oil recovery: an exciting new EOR opportunity for Alaska's north slope. In: Paper presented at the SPE western regional meeting, Irvine, California, 30 March–1 April. https://doi.org/10.2118/93903-ms

Merey Ş (2019) Analysis of the effect of experimental adsorption uncertainty on CH4 production and CO_2 sequestration in Dadas shale gas reservoir by numerical simulations. J Petrol Sci Eng 178:1051–1066. https://doi.org/10.1016/j.petrol.2019.04.022

Myshakin EM, Singh H, Sanguinito S, Bromhal G, Goodman AL (2019) Flow regimes and storage efficiency of CO_2 injected into depleted shale reservoirs. Fuel 246:169–177. https://doi.org/10.1016/j.fuel.2019.02.095

Naderi S, Simjoo M (2019) Numerical study of low salinity water alternating CO_2 injection for enhancing oil recovery in a sandstone reservoir: coupled geochemical and fluid flow modeling. J Petrol Sci Eng 173:279–286. https://doi.org/10.1016/j.petrol.2018.10.009

Nowrouzi I, Manshad AK, Mohammadi AH (2019) Effects of dissolved carbon dioxide and ions in water on the dynamic interfacial tension of water and oil in the process of carbonated smart water injection into oil reservoirs. Fuel 243:569–578. https://doi.org/10.1016/j.fuel.2019.01.069

Patel PD, Christman PG, Gardner JW (1987) Investigation of unexpectedly low field-observed fluid mobilities during some CO_2 Tertiary floods. SPE Reserv Eng 2(04):507–513. https://doi.org/10.2118/14308-PA

Peck WD, Azzolina NA, Ge J, Bosshart NW, Burton-Kelly ME, Gorecki CD, Gorz AJ, Ayash SC, Nakles DV, Melzer LS (2018) Quantifying CO_2 storage efficiency factors in hydrocarbon reservoirs: a detailed look at CO_2 enhanced oil recovery. Int J Greenhouse Gas Control 69:41–51. https://doi.org/10.1016/j.ijggc.2017.12.005

Pranesh V (2018) Subsurface CO_2 storage estimation in Bakken tight oil and Eagle Ford shale gas condensate reservoirs by retention mechanism. Fuel 215:580–591. https://doi.org/10.1016/j.fuel.2017.11.049

Rani S, Padmanabhan E, Prusty BK (2019) Review of gas adsorption in shales for enhanced methane recovery and CO_2 storage. J Petrol Sci Eng 175:634–643. https://doi.org/10.1016/j.petrol.2018.12.081

RezaeiDoust A, Puntervold T, Strand S, Austad T (2009) Smart water as wettability modifier in carbonate and sandstone: a discussion of similarities/differences in the chemical mechanisms. Energy Fuels 23(9):4479–4485. https://doi.org/10.1021/ef900185q

Seyyedi M, Mahzari P, Sohrabi M (2018) A comparative study of oil compositional variations during CO_2 and carbonated water injection scenarios for EOR. J Petrol Sci Eng 164:685–695. https://doi.org/10.1016/j.petrol.2018.01.029

Shakiba M, Ayatollahi S, Riazi M (2016) Investigation of oil recovery and CO_2 storage during secondary and tertiary injection of carbonated water in an Iranian carbonate oil reservoir. J Petrol Sci Eng 137:134–143. https://doi.org/10.1016/j.petrol.2015.11.020

Shu G, Dong M, Chen S, Luo P (2014) Improvement of CO_2 EOR performance in water-wet reservoirs by adding active carbonated water. J Petrol Sci Eng 121:142–148. https://doi.org/10.1016/j.petrol.2014.07.001

Snippe J, Berg S, Ganga K, Brussee N, Gdanski R (2020) Experimental and numerical investigation of wormholing during CO_2 storage and water alternating gas injection. Int J Greenhouse Gas Control 94:102901. https://doi.org/10.1016/j.ijggc.2019.102901

Tang G-Q, Morrow NR (1999) Influence of brine composition and fines migration on crude oil/brine/rock interactions and oil recovery. J Petrol Sci Eng 24(2):99–111. https://doi.org/10.1016/S0920-4105(99)00034-0

Teklu TW, Alameri W, Graves RM, Kazemi H, Al-sumaiti AM (2014) Low-salinity water-alternating-CO_2 flooding enhanced oil recovery: theory and experiments. In: Paper presented at the Abu Dhabi international petroleum exhibition and conference, Abu Dhabi, UAE, 11 Nov 2014. https://doi.org/10.2118/171767-ms

Teklu TW, Alameri W, Graves RM, Kazemi H, AlSumaiti AM (2016) Low-salinity water-alternating-CO_2 EOR. J Petrol Sci Eng 142:101–118. https://doi.org/10.1016/j.petrol.2016.01.031

Wilson A (2015) CO_2 low-salinity water alternating gas: a promising new approach for EOR. J Petrol Technol 67(01):84–86. https://doi.org/10.2118/0115-0084-JPT

Wu M, Ding M, Yao J, Li C, Li X, Zhu J (2019) Development of a multi-continuum quadruple porosity model to estimate CO_2 storage capacity and CO_2 enhanced shale gas recovery. J Petrol Sci Eng 178:964–974. https://doi.org/10.1016/j.petrol.2019.03.077

Zhang L, Ren B, Huang H, Li Y, Ren S, Chen G, Zhang H (2015) CO_2 EOR and storage in Jilin oilfield China: monitoring program and preliminary results. J Petrol Sci Eng 125:1–12. https://doi.org/10.1016/j.petrol.2014.11.005

Zhang P, Tweheyo MT, Austad T (2007) Wettability alteration and improved oil recovery by spontaneous imbibition of seawater into chalk: impact of the potential determining ions Ca^{2+}, Mg^{2+}, and SO_4^{2-}. Colloids Surf A 301(1):199–208. https://doi.org/10.1016/j.colsurfa.2006.12.058

Zhou J, Liu M, Xian X, Jiang Y, Liu Q, Wang X (2019) Measurements and modelling of CH_4 and CO_2 adsorption behaviors on shales: implication for CO_2 enhanced shale gas recovery. Fuel 251:293–306. https://doi.org/10.1016/j.fuel.2019.04.041

Zhou J, Xie S, Jiang Y, Xian X, Liu Q, Lu Z, Lyu Q (2018) Influence of supercritical CO_2 exposure on CH_4 and CO_2 adsorption behaviors of shale: implications for CO_2 sequestration. Energy Fuels 32(5):6073–6089. https://doi.org/10.1021/acs.energyfuels.8b00551

The manufacturer's authorised representative in the EU is Springer Nature Customer Service Centre GmbH, Europaplatz 3, 69115 Heidelberg, Germany. If you have any concerns regarding our products, please contact ProductSafety@springernature.com

Printed and bound by CPI Group (UK) Ltd, Croydon, CR0 4YY
25/09/2026
02222901-0013